T0107379

PUBLICATIONS OF THE ISRAEL ACADEMY

OF SCIENCES AND HUMANITIES

SECTION OF SCIENCES

———

FAUNA PALAESTINA

ARACHNIDA III: ARANEAE: THERIDIIDAE

FAUNA PALAESTINA

Editorial Board
THE FAUNA ET FLORA PALAESTINA COMMITTEE

Series Editor
F.D. POR

Executive and Science Editor
ILANA FERBER

כתבי האקדמיה הלאומית הישראלית למדעים

החטיבה למדעי-הטבע

החי של ארץ-ישראל

עכבישנים 3 : עכבישאים : כדורניים

(ARANEAE: THERIDIIDAE)

מאת

גרשׂם לוי

ירושלים תשנ״ח

כל הזכויות שמורות לאקדמיה הלאומית הישראלית למדעים

נדפס בישראל

תשנ״ח

To Rina Levy for her patience and constant encouragement

Photo: P. Amitai

Latrodectus tredecimguttatus, male.

Photo: P. Amitai

Steatoda paykulliana, female.

Latrodectus tredecimguttatus, female.

Photo: P. Amitai

FAUNA PALAESTINA · ARACHNIDA III

ARANEAE : THERIDIIDAE

by

GERSHOM LEVY

Jerusalem 1998

The Israel Academy of Sciences and Humanities

© The Israel Academy of Sciences and Humanities, 1998

Author's Address:
Department of Evolution, Systematics and Ecology
The Hebrew University of Jerusalem
91904 Jerusalem, Israel

ISBN 965–208–013–6
ISBN 965–208–133–7

Printed in Israel
Typeset at 'Graphit' Press Ltd., Jerusalem
Printed by Keterpress Enterprises Ltd., Jerusalem

CONTENTS

PREFACE

The theridiid spiders are well represented in Israel. Despite the small size of this country there are many genera and species, approximately the numbers recorded from many larger European countries which have been well investigated for a considerably longer period of time. The theridiid fauna of Israel is still not exhausted and additional new species, e.g., *Theridion* should be expected. I examined all the Theridiidae ever described from Israel and adjacent countries and many from other faunae as well. Twenty species, some with erroneus identifications were formerly reported from this area; emendations are listed in the Appendix at the end. Sixteen additional species including three genera represent new records of mainly African and south European species previously unknown from Israel. Many from both batches above were named more than one hundred years ago and have never been adequately described or illustrated. In addition 25 species new to science have been recognized. Former results and full documentation were published in a series by Levy & Amitai (1979–1983) and by Levy (1985a, 1991). The material is desposited in the Arachnological Collection of the Hebrew University of Jerusalem. The format of the description and morphological terms follow those used in the *Fauna Palaestina* publication on the Thomisidae by Levy (1985b).*

*   [The Chapter on 'Spider Morphology' is reprinted here from Levy (1985b: 6–12), for the reader's convenience.]

ACKNOWLEDGEMENTS

I am deeply indebted to Professor Herbert W. Levi, Museum of Comparative Zoology, Harvard University, USA, for the very close cooperation over a quarter of a century. His numerous revisions of the Theridiidae worldwide are followed here. Sincere thanks for the loan of specimens are due to Mr M. Hubert, Muséum National d'Histoire Naturelle, Paris; Dr M.W.R. de V. Graham and Mr E. Taylor, Hope Entomological Collections, University Museum, Oxford; Mr P.D. Hillyard and Mr F.R. Wanless, British Museum (Natural History), London; Dr M. Grasshoff, Forschungsinstitut Senckenberg, Frankfurt a.M.; Dr J. Gruber, Naturhistorisches Museum, Vienna; Dr T. Kronestedt, Naturhistoriska Riksmuseet, Stockholm; Dr J. Terhivuo and Dr L. Koli, Zoological Museum, Helsinki; Dr G. Arbocco, Museo Civico di Storia Naturale "G. Doria", Genova.

I am much obliged to Mr P. Amitai for the following drawings: 1, 2, 6–10, 17, 18, 21, 22, 29, 30, 44, 69, 77, 87–89, 96, 97, 104, 116, 123, 124, 131, 134, 135, 144, 152–154, 163, 165–171, 179, 180, 184–186, 191–195, 203, 204, 207, 213, 217, 218, 230–232, 238, 248, 252–255, 269, 270, 277–279, 289, 296, 297, 304, 305, 316, 337, 343, 355, 358, 380, 390, 401; all others were prepared by the author.

We are grateful to the publishers of the following journals for permission to reproduce figures from papers by G. Levy & P. Amitai and by G. Levy: The Zoological Society of London, *Journal of Zoology* (1982, 196:81–131; 1985, 207:87–123); The Linnean Society of London, *Zoological Journal of the Linnean Society* (1981, 72:43–67; 1983, 77:39–63); The Norwegian Academy of Science and Letters and The Royal Swedish Academy of Sciences, *Zoologica Scripta* (1982, 11:13–30); The British Arachnological Society, *Bulletin of the British Arachnological Society* (1981, 5:177–188; 1991, 8:227–232); The Weizmann Science Press, *Israel Journal of Zoology* (1979, 28:114–130). The study was supported by grants from the United States–Israel Binational Science Foundation (BSF), Jerusalem, and the *Fauna et Flora Palaestina* Committee of the Israel Academy of Sciences and Humanities.

INTRODUCTION

GENERAL BIOLOGY AND ZOOGEOGRAPHY

Generally, theridiids with their spherical opisthosoma, the usually dark anterior-median eyes and the legs devoid of spinose setae, have a characteristic appearance. Known deviations are *Theridion* species with opisthosomal humps, *Episinus* with the enlarged opisthosoma or *Argyrodes* with their triangular to extremely elongated opisthosoma. The marked sexual dimorphism often encountered adds another different habitus. Such are the circular coriaceous scutella, unique among Israeli theridiids, which are found on the opisthosomal back of the male of *Crustulina conspicua*, the exceptionally raised prosoma of *Dipoena* and *Argyrodes* males or the immensely enlarged chelicerae of male *Enoplognatha*.

Many theridiids are colourful, displaying conspicuous patterns of red or yellow markings or silver spots on a blackish background. Some are shiny black or brown or white with or without bands, streakes or spots in a different hue. In addition, marked colour variation in a population is not uncommon.

All theridiids of Israel have eight eyes. Some like *Coscinida*, *Dipoena* and *Euryopis* have relatively very large ones. Since most theridiids are web weavers sensitivity to vibrations and chemoreception rather than eyesight should be considered of primary importance.

With the exception of *Euryopis* species which hunt while foraging along twigs or on the ground, and *Argyrodes* which scavenge for food in various host webs, all theridiids of Israel whose biology is known are sedentary spiders, sit-and-wait predators that hang upside-down in their webs. In *Episinus* the web is reduced merely to a few fine vertical threads attached to the ground and held apart by the spider's legs. Usually, however, the webs are a network of strong threads of which a few are sticky; on touching, these are easily recognized in the field by experienced collectors. Irregular webs with an intersecting mass of threads sometimes known as cobwebs, are found in *Crustulina* or many *Steatoda* species. There are, however, some *Steatoda* that construct an expansive sheet suspended beneath an overhanging cliff or a plant and anchored below by a scaffold work of threads that are sticky on their lower ends.

Webs of *Theridion* being built out of very fine threads may be hardly visible while those of *Latrodectus* are often dusty and heavily interwoven with carcasses, dry leaves and pebbles. Silk walls mixed with pieces of earth are also found in the spherical retreats of certain *Enoplognatha* attached to the underside of stones. The webs of Widow spiders, *Latrodectus*, consist of a cone-shaped retreat connected to a horizontal capture platform. Entangled prey is enswathed by silk flung over by the spider's combed hind legs. With the prey being partly immobilized the spider usually bites its

3

legs in several lightning attacks. When the poison begins to take effect the spider gets closer and wraps a thick blanket of silk around the prey. In webs above the ground the prey might be hauled slowly upwards sometimes (e.g., a beetle, a scorpion or occasionally a lizard) weighing many times the weight of the spider.

Venoms of theridiids are toxic to invertebrates and in certain *Steatoda* species and some notorius Widows may be virulent to vertebrates including mammals. The toxicity of *Latrodectus* venom to mammals is considered to be due to proteins with neurotoxic action. Only the female spider has a significant amount of venom capable of harming mammals. Theridiids do not chew their prey; they bite it and suck it dry. Consequently residues of carcasses can be identified and the diet analyzed and information can thus be gained on occurrence and habits of some faunal components. The predominant prey of theridiids are insects. Some *Steatoda* and *Latrodectus* species prey heavily on ants, and *Euryopis* and *Dipoena* apparently specialize as ant predators. None of the theridiids of Israel are known to build communal webs or display any degree of social organization. The sexes are easily distinguished. The external epigynal plate of the female often has hardly any distinctive features but inner spermathecal organs exhibit characteristic shapes of specific diagnostic importance. Their complexity of structure and duct winding is sometimes difficult to trace. Considerable size differences between the sexes are seen mainly in the larger theridiids like *Steatoda* and particularly in *Latrodectus*. Stridulating organs, where present, consist of ridges posteriorly on the carapace which are rubbed against a row of denticles on the fore part of the opisthosoma. Courting is induced by the male plucking the female's web. Males may accidentally be devoured by the female. In webs of certain *Latrodectus* species, adult males were always found side by side with immature females.

Egg sacs are usually suspended in the web while the female spider abides in close proximity. Fifteen or more egg sacs can be laid by a single female in some *Steatoda* species. The number of eggs in a sac varies: from a few in *Dipoena* or *Crustulina* to several hundred in *Steatoda* and *Latrodectus*. There are different kinds of egg sacs: some are nearly transparent, others are formed of fluffy silk or coated with earth particles and others are tough coated and papery e.g., *Latrodectus* and *Argyrodes*. The duration of time from oviposition to emergence from the egg sac varies and may take up to several weeks. The number of moults and the length of the developmental period up to maturity likewise vary considerably. Theridiids in Israel with known biology show an annual life-cycle in nature. The life span of adult males in the field is usually of only a few weeks while the females may live for several months. Under laboratory conditions females of some species of *Steatoda* and *Latrodectus* have been kept alive for over two years.

Theridiids in Israel have been collected from the Dead Sea area, 400 m below sea level to Mount Hermon at heights above 2000 m, and from the eremic southern Negev to the relatively humid coastal plain of the Mediterranean. There are species which live on the ground, in litter, under stones, inside rock crevices, in grass, bushes and up in the trees. In Israel, species of *Theridion* for example can be found in all strata of vegetation and habitats. *Coscinida* or *Dipoena* speices on the other hand, are rarely encountered.

Evidently, apart from occasional, somewhat dense populations of *Latrodectus* and small aggregations of some *Steatoda* species under logs or inside animal breeding facilities, no great number of theridiids are ever encountered in Israel.

The Theridiidae is one of the larger spider families. About 2000 species in the world in about 45 genera are placed according to the generally accepted classification resulting from the many extensive revisions conducted by Herbert W. Levi. Many common spiders throughout the world are members of this family. Twelve genera with a total of 61 species are known at present from Israel. Only a few are Holarctic, e.g., *Crustulina sticta*, *Steatoda albomaculata*, *S. triangulosa*, *Theridion melanurum* and *T. simile*. Some, e.g., *Theridion melanostictum*, *Steatoda erigoniformis* and *Latrodectus* aff. *hesperus* are distributed worldwide in nature, possibly not as the result of human introduction. Others, however, like *Theridion rufipes*, *Steatoda grossa* and *Latrodectus geometricus* are synantropic spiders and were carried by man around the world.

The majority of the Israeli theridiids are Mediterranean elements. These are species that may occur in southern Europe, North Africa and the Near East. Some North African species may belong to the Palaeo-eremic fauna — the fauna of the Old World Desert belt extending south and east of the Mediterranean region. Such eremic species may reach the southern parts of Israel. With the paucity of information it cannot be settled yet whether species found along the Jordan Rift represent relicts of an ancient tropical fauna, or a northward influx of Ethiopian elements, or represent Palaeo-eremic species which have adapted to the special environmental conditions of oases in the Rift Valley. Endemic species may occur in the arid zones along the Rift Valley. However, conclusive evidence on endemics cannot be gained unless better knowledge is acquired on the spider fauna of the adjacent countries.

Reprinted from Levy (1985b)

SPIDER MORPHOLOGY

The spider body consists of two parts, the anterior *prosoma* (= cephalothorax) and the posterior *opisthosoma* (Figs. 1, 2). The two portions are connected by a narrow stalk, the pedicel, enabling free movement of each portion in all directions. The prosoma combines the head and thorax functions; it contains the highly condensed nervous system and bears the eyes, mouthparts, pedipalps and walking legs. The opisthosoma bears the openings of the respiratory, reproductive and digestive systems, and also the spinnerets.

Prosoma: It is covered above by an unsegmented *carapace* (Fig. 1) and below by a large plate, the *sternum* (Fig. 2). Attached to the anterior edge of the sternum, usually delimited by a groove, is a median platelet serving as a *labium* (Figs. 2, 4). The sternum is often indented opposite the coxa of each leg. The *eyes* are variously distributed over the carapace (Figs. 1, 3). Their disposition is characteristic in different families and is used extensively in taxonomy. They are often arranged in two

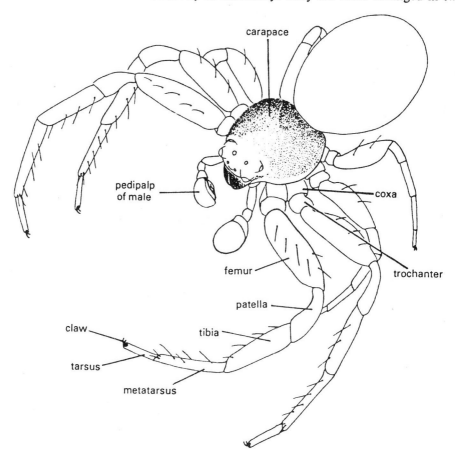

* Fig. 1: Male spider, dorsal view

6

* [Figures 1–10 on 'Spider morphology' are reproduced here from *Fauna Palaestina: Arachnida II: Araneae: Thomisidae* (Levy, 1985b).]

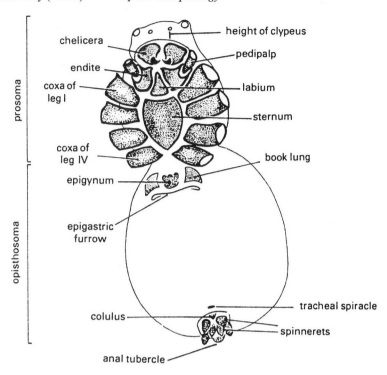

* Fig. 2: Female spider, ventral surface
illustrating main external features (legs omitted)

rows, the *anterior row* and the *posterior row*, the inner two eyes of each row being the *medians*, the outer eyes — the *laterals* (Fig. 3). This terminology is retained through-out even though in some spiders the specific term does not actually describe the location of a particular pair of eyes. The rows of eyes are frequently curved. When the lateral eyes of a row are nearer to the front end of the carapace than are the median eyes, the row is termed *procurved*, and the opposite situation is termed *recurved* (Fig. 3). In specifying the curvature it is assumed that the eyes are viewed from vertically above; the actual curve of the anterior row is checked from directly in front. The area enclosed by the anterior-median and posterior-median eyes is termed the *median ocular quadrangle* and is useful taxonomically (Fig. 3). When both dark and light-coloured eyes are present in the same spider they are referred to as *hetero-geneous*, whereas if the eyes are all alike they are called *homogeneous*. The region between the anterior eyes and the front margin of the carapace is termed the *clypeus*. The *height* of the clypeus refers to the distance from the front edge of the carapace to the eyes nearest that edge (Fig. 2).

Mouthparts: A pair of *chelicerae* are positioned above the mouth. Each chelicera consists of a stout basal segment and a distal fang (Fig. 4). The fang folds onto the basal segment along a groove whose edges are frequently provided with rows of small teeth. The teeth are often of importance in identification. Those on the outer margin

7

* [Figures 1–10 on 'Spider morphology' are reproduced here from *Fauna Palaestina: Arach-nida II: Araneae: Thomisidae* (Levy, 1985b).]

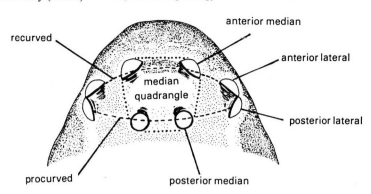

* Fig. 3: Diagram of eyes' arrangement on dorsal, anterior part of carapace

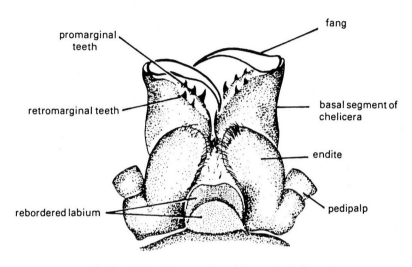

* Fig. 4: Mouthparts of a spider, ventral view

(= anterior or upper row) are termed *promarginal* and those on the inner margin — *retromarginal* (Fig. 4); the latter may be visible from below. Close to the tip of the fang is the opening of the poison gland. This gland either lies entirely in the chelicerae or, as in most spiders, extends into the prosoma.

Pedipalpi: Situated behind the chelicerae, the pedipalps, usually referred to as *palps*, are leg-like in females, or modified in mature males to contain the copulatory organs (Fig. 1). In most spiders their coxae are greatly enlarged on the inner side to form flattened *endites* enclosing the preoral cavity (Figs. 2, 4). In the male palp, often the tibia, sometimes also the patella and even the femur, may be enlarged and bear *apophyses* and special spines which are of great taxonomic importance (Figs 5, 6). In the majority of spiders the distal part of the male palp, the *cymbium*, is hollowed out to accommodate the functional components of the copulatory organ. The palpal organ of the male palp stores the sperm derived from the genital orifice situated below the anterior part of the opisthosoma, and is used to transfer the sperm into the

8

* [Figures 1–10 on 'Spider morphology' are reproduced here from *Fauna Palaestina: Arachnida II: Araneae: Thomisidae* (Levy, 1985b).]

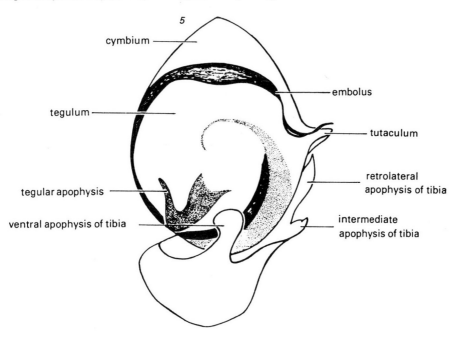

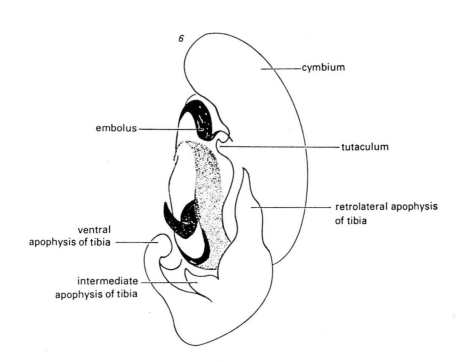

* Figs. 5–6: Diagram of male palpal organ; left palpus
5. ventral view; 6. lateral view

* [Figures 1–10 on 'Spider morphology' are reproduced here from *Fauna Palaestina: Arachnida II: Araneae: Thomisidae* (Levy, 1985b).]

female during mating. In many families, the palpal organ is expanded during mating, unfolding the intromittent portion, the *embolus*, and its various sclerites and apophyses (Figs. 5, 6). In the male, the complex structure of the palpal organ provides the character of greatest systematic value, particularly at the species level. The palpal organ is illustrated for each species, usually in its unfolded position.

Legs: The four pairs of legs are attached below the edges of the carapace (Fig. 1). Each leg consists of seven segments terminating in either two or three claws. The claws are frequently toothed (pectinate). The legs are usually covered with hairs and often bear spines and bristles. Many spiders have bundles of strong hairs termed *claw tufts* beneath the claws (Fig. 7). In some groups of spiders, the ventral surface of the leg's distal segment, the tarsus and sometimes also the metatarsus, are clothed with a dense brush of short, stiff hairs, the *scopula* (Fig. 7). Spiders possessing a specialized spinning organ, called the *cribellum*, on the opisthosoma also have a comb-like series of small, curved bristles, termed the *calamistrum,* on the dorsal surface of the metatarsi of the hind legs (Figs. 8, 9). The calamistrum serves to card very fine threads of silk out of the cribellum.

Opisthosoma: The opisthosoma lacks visible external segmentation in all spiders of the Middle East; it is most commonly oval in form but it may be modified in many ways. Some of these modifications may take the form of protuberances or a long tail. Many spiders have a coloured pattern on their back and venter, but it may vary greatly within species; bright colours usually do not persist long in alcohol. The dorsal surface of the opisthosoma frequently bears a series of depressed spots of varying size, indicating the points of attachment of internal muscles to the body wall.

In all spiders there is a distinct transverse groove on the anterior half of the ventral surface of the opisthosoma, which is called the *epigastric furrow* (Figs. 2, 10). In the mature spider, the genital orifice opens in the mid-region of this furrow. In the great majority of mature female spiders a sclerotized plate is located in the region in front of the epigastric furrow and is associated with the genital opening, the *epigynum* (Figs. 2, 10). Most mature female spiders have external copulatory orifices on the epigynal surface, connected via ducts to internal seminal receptacles, the *spermathecae* (Fig. 10). Sperm is stored in the spermathecae and passed via ducts to the oviduct

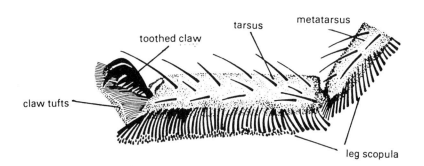

* Fig. 7: Terminal joints of walking leg with scopula and claw tufts

10

* [Figures 1–10 on 'Spider morphology' are reproduced here from *Fauna Palaestina: Arachnida II: Araneae: Thomisidae* (Levy, 1985b).]

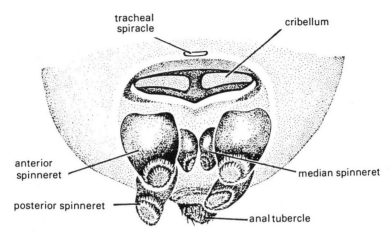

* Fig. 8: Anal part of spider showing cribellum and spinnerets

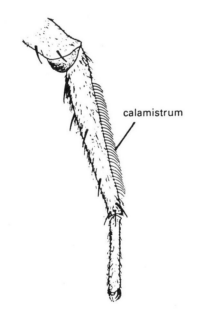

* Fig. 9: Fourth leg, portion showing bristles on
metatarsus forming the calamistrum

11

* [Figures 1–10 on 'Spider morphology' are reproduced here from *Fauna Palaestina: Arachnida II: Araneae: Thomisidae* (Levy, 1985b).]

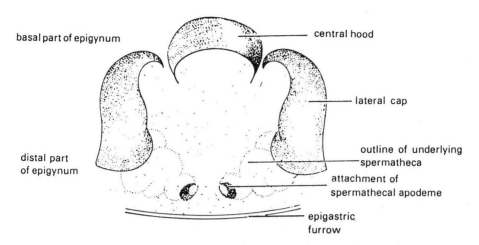

basal part of epigynum

central hood

lateral cap

distal part
of epigynum

outline of underlying
spermatheca

attachment of
spermathecal apodeme

epigastric
furrow

* Fig. 10: Diagram of epigynal plate of female, ventral view

to fertilize eggs as they are laid. Some spider families have no separate, external copulatory pores and the spermathecae open only into the oviduct; the intromittent organ of the male palpus in these spiders is inserted directly into the genital orifice. The internal spermathecal organs may be partially discernible through the epigynal integument; dark circles, indicating attachment points of internal apodemes, are sometimes visible on the outside (Fig. 10). The epigynum, spermathecae and also the ducts show an infinite variety of form and, like the male palpal organ, are very important in the identification of species.

Spiders may be equipped with either one or two pairs of respiratory organs which open on the ventral side of the opisthosoma (Fig. 2). In the different families, several combinations of lamellate organs ('book lungs') and tracheae of various types and origins are to be found. A pair of book lungs covered by light-coloured plates is situated in front of the epigastric furrow in most spiders, and a single slit of the fused tracheae opens posteriorly near the base of the spinnerets.

At the morphologically posterior end of the opisthosoma is the opening of the digestive system, which is situated on a more or less distinct *anal tubercle* (Figs. 2, 8). Below this are the *spinnerets* which in all spiders of the Middle East may number one, two or three pairs. The spinnerets are jointed appendages — some reduced to a single joint — with a terminal, membranous portion studded with many minute tubes through which liquid silk is extruded. The disposition, relative length and number of joints of the spinnerets are useful in classification. Usually, the medians are the smallest and are often hidden by the others. In a number of families there is a sieve-like plate, the *cribellum*, in front of the spinnerets, from which a special type of silk is carded (Fig. 8). Many of the spiders lacking a cribellum have a vestigial, conical lobe, termed the *colulus*, between the bases of the anterior spinnerets (Fig. 2). Silk glands with external, isolated pores are also found anteriorly on the opisthosoma in many male spiders.

12

* [Figures 1–10 on 'Spider morphology' are reproduced here from *Fauna Palaestina: Arachnida II: Araneae: Thomisidae* (Levy, 1985b).]

SYSTEMATIC PART

Family THERIDIIDAE Sundevall, 1833
Conspectus Arachnidum, Londini Gothorum, p. 15

Diagnosis: Small to medium-size spiders, about 1–15 mm in body length, most less than 6 mm; gravid females of some Widow species may attain a length of 20 mm.

A row of serrated setae is present (except in *Argyrodes*) on fourth tarsus. Male palpus bears a paracymbial-hook behind bulb, inside cymbium or on edge of cymbium but not on base. Metatarsi generally longer than tarsi. Cheliceral teeth absent or only few in number. Distal margin of labium not swollen (rebordered).

As in related families like the Araneidae and Linyphiidae, the Theridiidae have three claws, the tarsi are without trichobothria, a colulus may be present or absent or replaced by two setae, the anterior and posterior spinnerets are short and conical and cover the small middle pair, and with the anal tubercle form a circle. There is an epigynum in the female and a complex palpus in the male but without tibial apophyses; a tracheal spiracle is near the spinnerets.

Key to the Genera of *Theridiidae* in Israel
based on key to Holarctic theridiid genera by H.W. Levi

1. Colulus present between anterior spinnerets   2
- Colulus absent or only two setae present between spinnerets   7
2. Prosoma covered throughout with tubercles; cymbium in palpus of male with mesal projection appears bilobed (Fig. 12); opisthosoma spherical   **Crustulina**
- Prosoma without tubercles or only slightly rugose in some males but palpus then without mesal projection; if palpus bilobed then opisthosoma triangular   3
3. Lateral eyes separated by their diameter or more; chelicerae without teeth   **Latrodectus**
- Lateral eyes touching or slightly separated at most; chelicerae with teeth   4
4. Male with raised eye region or frontal projection, or with deep recess below eyes; opisthosoma of female triangular or much elongated with humps extending beyond spinnerets   **Argyrodes**
- Male eye region not so modified; female opisthosoma oval to spherical   5
5. Chelicerae of female without retromarginal teeth; male with robust, widened fangs (if fangs of ordinary size then chelicerae with only one small promarginal tooth)   **Steatoda**
- Chelicerae of female with at least one retromarginal tooth; male with much enlarged chelicerae and teeth or with ordinary chelicerae armed with several promarginal teeth   6
6. Male with disproportionally large chelicerae and much enlarged teeth; cymbium in palpus of male with paracymbial-hook on ectal margin; dark opisthosoma of female with a conspicuous pattern   **Enoplognatha**
- Male with small, ordinary chelicerae; paracymbial-hook hidden in back of palpus; greyish opisthosoma of female without distinct pattern   (in part) **Anelosimus**

13

7. Colulus represented by two setae   8
- Colulus absent   11
8. Opisthosoma longer than wide, widest behind middle and may have humps posteriorly
     Episinus
- Opisthosoma not so modified   9
9. Chelicerae with teeth, fangs short   (in part) **Anelosimus**
- Chelicerae without teeth, fangs long   10
10. Opisthosoma triangular, widest in front and pointed behind; stout prosoma of male without clypeal recess   **Euryopis**
- Opisthosoma globular; high prosoma of male with eye region projecting above concave clypeus   **Dipoena**
11. Large eyes closely grouped in blackened area; male palpus with distinct hook on mesal margin of cymbium   **Coscinida**
- Eyes not closely grouped in blackened region; palpus without mesal hook   12
12. Opisthosoma higher than long with streaky pattern on sides   **Achaearanea**
- Opisthosoma spherical or slightly modified but not streaked   **Theridion**

Genus CRUSTULINA Menge, 1868
Schrift. Naturf. Gesell. Danzig, N.F. 2:168
Figs 1–7

Type-species: *Theridion guttatum* Wider, 1834

Spiders about 1.5 to 4.5 mm total length. Prosoma covered throughout by small tubercles (Figs 1, 2). Carapace longer than wide. Eyes about subequal in size. Anterior-median eyes about their diameter apart and distinctly closer to anterior-laterals (Fig. 3); all posterior eyes approximately their diameter apart or, posterior-medians slightly closer to posterior-laterals than to each other (Fig. 4). Clypeus in male markedly higher than in female. Chelicerae without retromarginal teeth (Fig. 5). Legs relatively short, first or fourth pair longest, third shortest. Opisthosoma almost spherical; pedicel in male sclerotized, in female sclerotization around pedicel sometimes incomplete. Colulus relatively large. Male may have large sclerotic patches on back of opisthosoma (Figs 6, 7).

Crustulina species construct an irregular, sparse web of fine, barely visible threads. They live inside rock fissures, under stones, in litter often in moist places, on moss or among water-dripping ferns. Being cryptic spiders, specimens are scarcely detected despite the conspicuous pattern on back. All species in Israel have a small, distinctive, white median spot on the venter of the opisthosoma. Up to 10 pink eggs are laid in a white spherical sac.

Crustulina species in Israel have chelicerae with one promarginal tooth and the palpus of the male has a distinct mesal process, thus appearing bilobed. Less than 20 species distributed throughout the world. None apparently found anywhere in abundance. Four species in Israel, one is known by the female only.

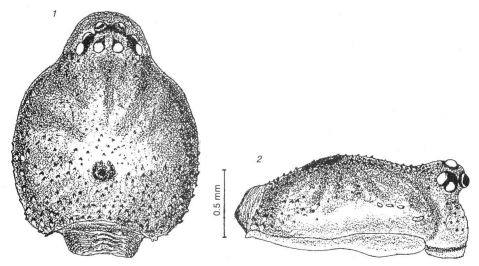

Figs 1–2: *Crustulina conspicua* (O.P.-Cambridge, 1872); male
1. carapace, dorsal view; 2. prosoma, lateral view

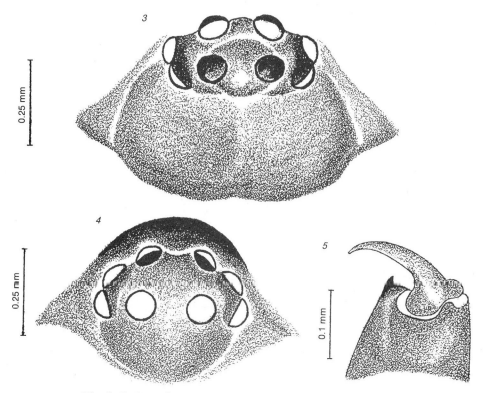

Figs 3–5: *Crustulina conspicua* (O.P.-Cambridge, 1872); female
3. carapace, frontal view; 4. carapace, dorsal view of anterior part;
5. tip of left chelicera, inner view

Key to the species of **Crustulina** in Israel

Males:

1. Back of opisthosoma with four large, circular, sclerotic impressions (Fig. 6)
   C. conspicua (O.P.-Cambridge)

   – Back of opisthosoma without large, conspicuous impressions   2

2. Falcated basal apophysis (BA) of palpus with accessory denticles on outer edges of blade (Figs 25, 26) and inner accessory apophysis with only one prong (IA; Fig. 25)
   C. sticta (O.P.-Cambridge)

   – Basal apophysis of palpus with accessory denticles placed along middle line of blade (Figs 33, 34) and inner accessory apophysis armed with two prongs (Fig. 33)
   C. scabripes Simon

Females:

1. Pattern on back of opisthosoma formed of separate white spots, distinct in particular posteriorly (Figs 17, 29).   2

   – Pattern in form of a network of mostly continuous white markings (Figs 9, 21)   3

2. Epigynum traversed centrally by a dark, raised fold and a nearly circular depression (Fig. 35).   **C. scabripes** Simon

   – Epigynum traversed by a fine, dark line with coned thickenings at ends (Fig. 19)
   C. hermonensis Levy & Amitai

3. Epigynum traversed centrally by a raised, sclerotized, bridge-like fold and a laterally extended depression (Fig. 27)   **C. sticta** (O.P.-Cambridge)

   – Epigynum traversed by a dark sclerotized line with thickened ends (Fig. 15)
   C. conspicua (O.P.-Cambridge)

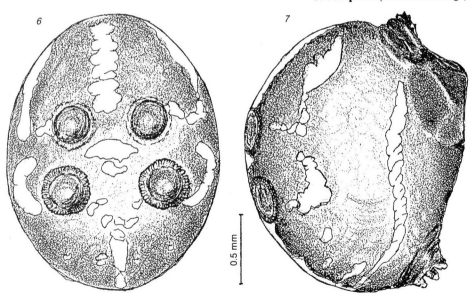

Figs 6–7: *Crustulina conspicua* (O.P.-Cambridge, 1872); male, opisthosoma
6. dorsal surface; 7. lateral view

Crustulina conspicua (O.P.-Cambridge, 1872)
Figs 1–16

Theridion conspicuum O.P.-Cambridge, 1872, *Proc. zool. Soc. Lond.*, p. 285, pl. 13, fig. 11.

Crustulina conspicua —. Simon, 1881, *Les Arachnides de France*, 5:160; Roewer, 1942, *Katalog der Araneae*, 1:398; Bonnet, 1956, *Bibliographia Araneorum*, 2(2):1252; Levy & Amitai, 1979, *Israel J. Zool.*, 28:117.

Length of male 3.3–4.2 mm, female 3.2–4.4 mm. Coloration of carapace deep brown, male with distinct median indentation (=fovea; Figs 1, 2). Sternum brown with black margins (Fig. 8). Legs yellow, sometimes slightly annulated. Opisthosoma dark brown, in male with a series of white markings on back extending over sides and four large, circular coriaceous impressions (Figs 6, 7), in female, with a network of partly continuous white markings covering back and sides (Figs 9, 10).

Male Palpus: Femur long and slender almost throughout (Fig. 11); patella distally enlarged, angular, with a slight apical recess (Fig. 12). Smooth, falciform basal apophysis at base of embolar division without accessory denticles; inner accessory apophysis absent (Figs 13, 14).

Female Epigynum: Shield shaped, slightly convex epigynal plate traversed proximally by a fine dark sclerotic line (Fig. 15); black, small, cone-like thickenings at ends of line with orifices turned obliquely towards epigastric furrow (Fig. 15). Inner spermathecae partly discernible occasionally through integument. Relatively short tubes of small globular spermathecae curve under a transparent, elaborate framework (Fig. 16).

Distribution: Israel, possibly Lebanon, Syria and northern Jordan.

Israel: Golan Heights (18) and Galilee (1) to Judean Hills (11) throughout.

Adult males are found mainly from March to May, and mature females from April to July. Oviposition of an egg sac with seven eggs was observed towards mid July in a female from the Golan Heights.

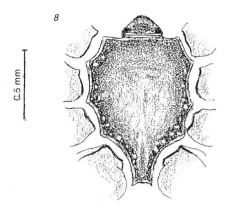

Fig. 8: *Crustulina conspicua* (O.P.-Cambridge, 1872); male, sternum, ventral surface

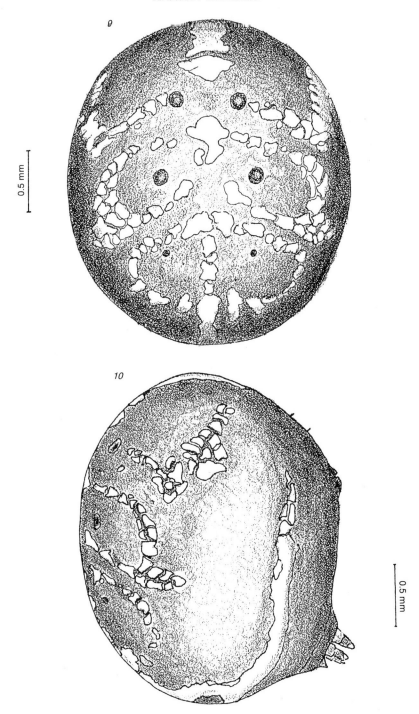

Figs. 9–10: *Crustulina conspicua* (O.P.-Cambridge, 1872); female, opisthosoma
9. dorsal surface; 10. lateral view

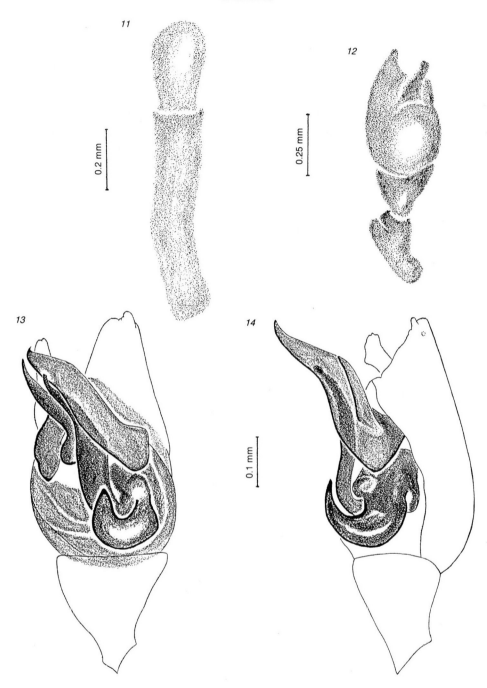

Figs 11–14: *Crustulina conspicua* (O.P.-Cambridge, 1872); male, left palpus
11. dorsal view of femur and patella; 12. mesal view of patella, tibia and cymbium;
13. ventral view; 14. retrolateral view

19

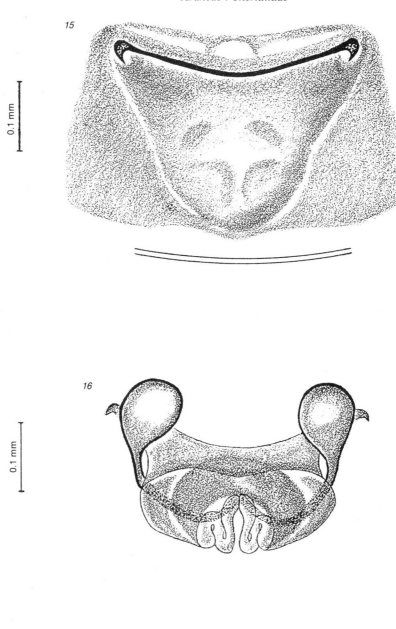

Figs 15–16: *Crustulina conspicua* (O.P.-Cambridge, 1872); female
15. epigynum; 16. inner spermathecae, dorsal view

Crustulina hermonensis Levy & Amitai, 1979
Figs 17–20

Crustulina hermonensis Levy & Amitai, 1979, *Israel J. Zool.*, 28:122.

Adult male unknown. Length of female 3.0–3.4 mm. Coloration of prosoma red brown. Legs yellowish-brown with a few light annulations. Opisthosoma dark with pattern of white extended markings which posteriorly turn into irregular separate spots (Figs 17, 18).

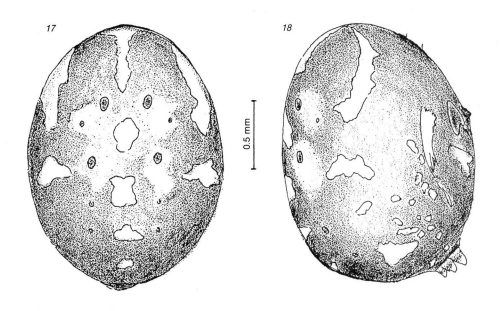

Figs 17–18: *Crustulina hermonensis* Levy & Amitai, 1979; female, opisthosoma
17. dorsal surface; 18. lateral view

Female Epigynum: Relatively large. Epigynal plate traversed proximally by fine shallow furrow marked by a dark sclerotized line (Fig. 19); dark, small, cup-like thickenings at ends of sclerotized line with orifices directed almost towards each other (Fig. 19). Inner organs are partly visible through integument. Tubes of globular spermathecae are fastened inside a partly sclerotized, complex structure (Fig. 20).
Distribution. Israel: Mt Hermon, 1800 m (19).
Adult females and young males were found under stones in October.

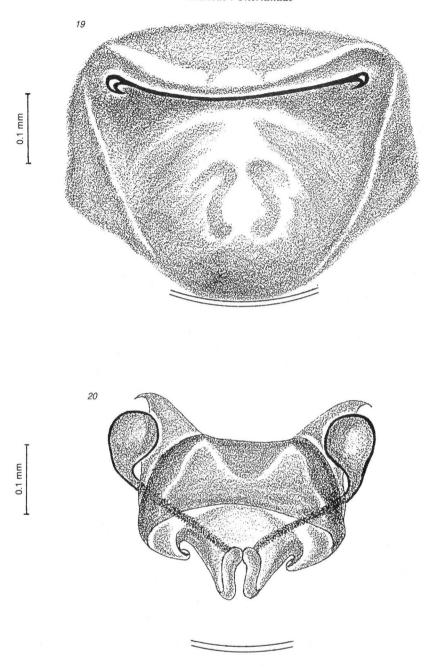

Figs 19–20: *Crustulina hermonensis* Levy & Amitai, 1979; female
19. epigynum; 20. inner spermathecae, dorsal view

Crustulina sticta (O.P.-Cambridge,1861)
Figs 21–28

Theridion stictum O.P.-Cambridge, 1861, *Ann. Mag. Nat. Hist.* London (3)7:432.

Crustulina sticta —. Simon, 1881, *Les Arachnides de France*, 5:158, Roewer, 1942, *Katalog der Araneae*, 1:399, Bonnet, 1956, *Bibliographia Araneorum*, 2(2):1256; Levi, 1957, *Bull. Mus. comp. Zool. Harv.*, 117:370; Levy & Amitai, 1979, *Israel J. Zool.*, 28:123.

Length of male 2.3 mm, female 2.8–3.5 mm. Carapace and sternum dark brown. Legs yellow and partly light brown. Opisthosoma in male whitish dorsally with a few dark patches in front and on sides, in female dark on back and sides with a network of more or less continuous white markings (Figs 21, 22).

Male Palpus: Femur and patella distally enlarged, claviform (Fig. 23); patella on inner view, angular (Fig. 24). Large, falciform basal apophysis (BA) armed with about six distinct, pointed accessory denticles on outer margin of blade (Figs 25, 26); sclerotized, styliformed inner accessory apophysis (IA) projects between base of embolar division and falcated apophysis (Figs 25, 26).

Female Epigynum: Dark, sclerotized fold with slightly raised cup-like corners extends across middle of plate (Fig. 27); membranous area proximally, above sclerotized fold surrounds a large narrow depression. Globular spermathecae rather small, their long tubes, twisting inside a transparent, sclerotic structure (Fig. 28).

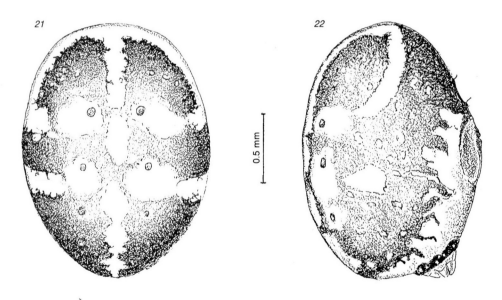

Figs 21–22: *Crustulina sticta* (O.P.-Cambridge, 1861); female, opisthosoma
21. dorsal view; 22. lateral view

23

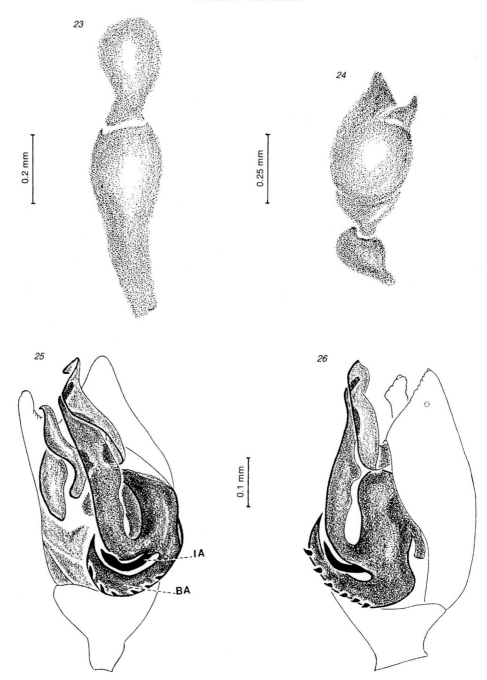

Figs 23–26: *Crustulina sticta* (O.P.-Cambridge, 1861); male, left palpus
23. dorsal view of femur and patella; 24. mesal view of patella, tibia and cymbium;
25. ventral view; BA – basal apophysis, IA – inner accessory apophysis;
26. retrolateral view

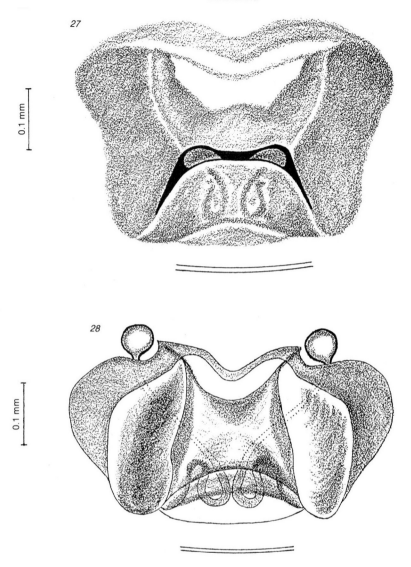

Figs 27–28: *Crustulina sticta* (O.P.-Cambridge, 1861); female
27. epigynum; 28. inner spermathecae, dorsal view

Distribution: Holarctic.

Israel: Upper Galilee (1), Carmel Ridge (3) and near Palmahim in the southern Coastal Plain (9).

An adult male was found in January, near Palmahim and adult females were collected in June, September and December on the Carmel and in the Galilee. All were found in litter in moist places.

25

Crustulina scabripes Simon, 1881
Figs 29–36

Crustulina scabripes Simon, 1881, *Les Arachnides de France*, 5:159; Roewer, 1942, *Katalog der Araneae*, 1:398; Bonnet, 1956, *Bibliographia Araneorum*, 2 (2):1255; Levy & Amitai, 1979, *Israel J. Zool.*, 28:126.

Length of male 3.1 mm; female as yet unknown from Israel. Coloration of male: carapace brown with fine dark stripes radiating from central deep fovea. Sternum almost black. Legs yellow with upperside of femora slightly darkened. Opisthosoma dark with white markings turning posteriorly into distinct spots. Opisthosomal pattern of female from French type material depicted (Figs 29, 30).

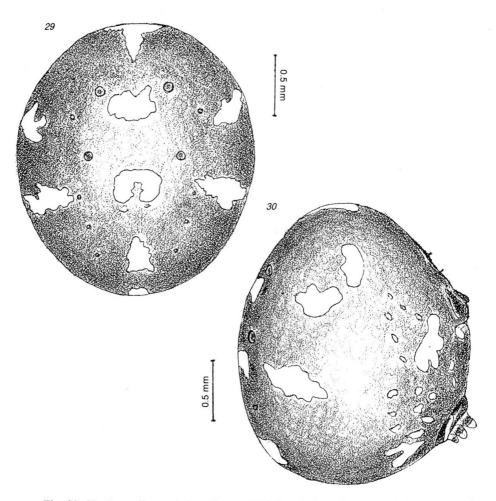

Figs 29–30: *Crustulina scabripes* Simon, 1881; female from France, opisthosoma
29. dorsal surface; 30. lateral view

Male Palpus: Femur and patella slightly enlarged distally (Fig. 31); patella, on inner view, angular (Fig. 32). Mesal process of cymbium armed with large, distinct warts (Fig. 32). Falciform basal apophysis armed with about five large, pointed accessory denticles along middle line of blade (Figs 33, 34); sclerotized, two-pronged inner accessory apophysis projects at base of embolar division (Fig. 33).

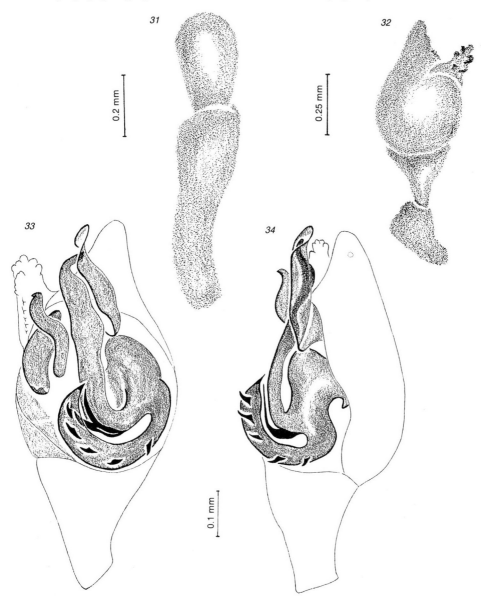

Figs 31–34: *Crustulina scabripes* Simon, 1881; male, left palpus
31. dorsal view of femur and patella; 32. mesal view of patella, tibia and cymbium;
33. ventral view; 34. retrolateral view

Female Epigynum: Drawings presented (Figs 35, 36) are of a female from France.
Distribution: Southern Europe, Algeria, Israel.
Israel: Golan Heights (18).
An adult male was found in April, females were all immature at that time of the year.

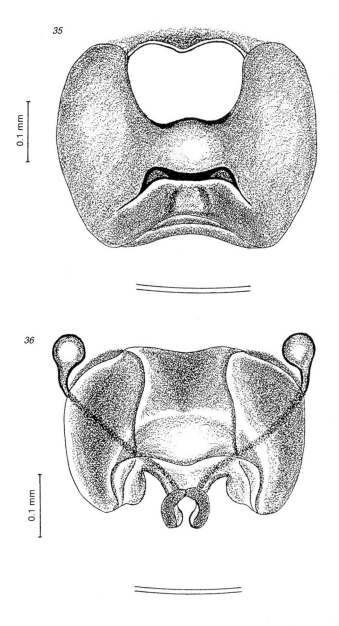

Figs 35–36: *Crustulina scabripes* Simon, 1881; female from France
35. epigynum; 36. inner spermathecae, dorsal view

Genus ENOPLOGNATHA Pavesi, 1880
Annali Mus. civ. Stor. nat. Genova 15:325
Figs 37–43

Type species: *Theridion mandibulare* Lucas, 1846

Theridiids of moderate size, about 3 to 6 mm total length. Carapace clearly longer than wide. Carapace of male sometimes slightly raised in front, and always with distinct stridulating ridges posteriorly (Figs 37, 38). Anterior and posterior eye rows almost straight (Figs 38–41); eyes about equal in size, but anterior-medians often the smallest. Anterior-lateral eyes and posterior-laterals usually touching. Posterior tip of sternum slender, tapering slightly between coxae of hind legs (Fig. 42). Chelicerae of male much enlarged (Figs 37, 53, 76), female with a few promarginal teeth and one retromarginal tooth (Fig. 43). First and fourth legs longest and about equal in length, but in male first often the longest; third legs shortest. Opisthosoma nearly oval (Figs 44, 52); male with a sclerotized, roughened area in front, above pedicel (Fig. 37). Colulus distinct, usually bearing a few setae.

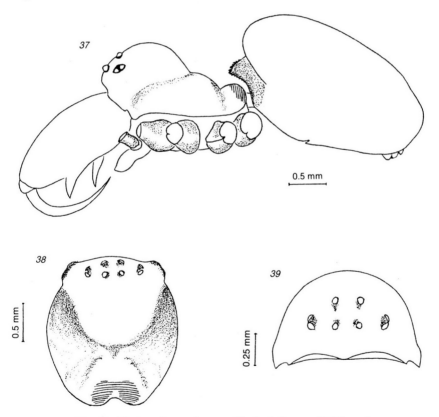

Figs 37–39: *Enoplognatha mandibularis* (Lucas, 1846); male
37. lateral view of spider (legs omitted); 38. carapace, dorsal view; 39. carapace, frontal view

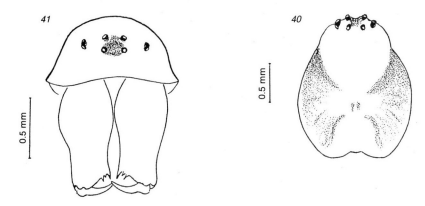

Figs 40–41: *Enoplognatha macrochelis* Levy & Amitai, 1981; female, carapace
40. dorsal view; 41. frontal view

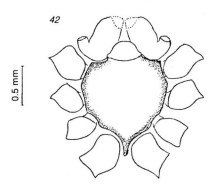

Fig. 42: *Enoplognatha mandibularis* (Lucas, 1846); male, sternum, ventral surface

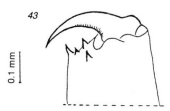

Fig. 43: *Enoplognatha macrochelis* Levy & Amitai, 1981; female,
tip of left chelicera, inner view

30

Enoplognatha species are usually found under stones or running on the ground in grass, seldom on shrubs. Their opisthosomal coloration is generally one of two kinds, a wide marking with leaf-like outline, termed folium, or an arrangement of white spots. Among those with the folium pattern there are species in which the folium is restricted to the dorsal surface, its margins generally visible from above; in other species, usually with a black, glossy hue, the folium extends well over the sides of the opisthosoma and its margins are marked by a conspicuous white fringe. The white-spotted pattern is more variable: some are grey to yellowish-brown, others are dark and the distribution of white spots is variable.

Males of *Enoplognatha* are noted for their enormous chelicerae, occasionally as long as their entire prosoma (Fig. 37). Immature males, up to the penultimate stage, have chelicerae of ordinary size like those of adult females. Females are often found clinging to the underside of a stone, together with one or more egg sacs. *Enoplognatha macrochelis* were sometimes found sitting inside a small, circular web, 1.5–2.0 cm in diameter, the silk walls being heavily mixed with pieces of earth. The egg sacs are yellowish-white, spherical, about 4 mm in diameter and contain numerous eggs. The eggs, usually pink-coloured, are visible through the sac wall.

Enoplognatha species may resemble each other closely and sometimes prove difficult to separate; some are also quite variable. About 50 species are reported from throughout the world, but many more are expected also from Israel. Five species are known from Israel.

Key to the species of **Enoplognatha** in Israel

1. Opisthosoma with a folium-like dorsal pattern (Fig. 44)   2
   – Opisthosoma with a white-spotted dorsal pattern (Fig. 69, 77)   4
2. Tibia of male palpus slender, much longer than bulb of palpus (Fig. 47). Posterior edge of epigynal plate of female projecting lip-like into epigastric furrow (Fig. 50); dark, thick sclerotized central structure of epigynum without discernible orifices (Fig. 50)

   E. mandibularis (Lucas)

   – Tibia of male palpus about as long as bulb; epigynum otherwise   3
3. Large, distal tooth (d, Fig. 45) of male chelicera without small, accessory denticles at base (Fig. 54). Slightly flattened orifice of central structure of epigynum touching posterior edge of epigynal plate (Fig. 58); large ducts of spermathecae form distinct loops (requiring dissection to reveal them) (Figs 59, 60)   **E. macrochelis** Levy & Amitai
   – Large, distal tooth of male chelicera armed with small, accessory denticles at base (Fig. 62). Partly divided orifice of central structure of epigynum placed at end of swelling extending to epigastric furrow (Fig. 66; from certain angles it looks like a horizontal figure 8); indistinct ducts of spermathecae fused to sides of median, sclerotized plate (Fig. 67)

   E. deserta Levy & Amitai
4. Thick, squat chelicera of male with a short, stout distal, stub-like denticle (Fig. 70). Black, heavy sclerotized epigynal plate bearing two large orifices (Fig. 74)

   E. parathoracica Levy & Amitai

   – Elongated chelicera of male with a long, pointed, distal tooth (Fig. 78). Epigynum traversed by dark, bridge-like central structure (Figs 82, 84)   **E. mediterranea** Levy & Amitai

Enoplognatha mandibularis (Lucas, 1846)

Figs 37–39, 42, 44–51

Theridion mandibulare Lucas, 1846, *Explor. scient. Algér.*, Zool. 1., p. 260, pl. 17 fig. 1.

Enoplognatha mandibularis —. Pavesi, 1880, *Annali Mus. civ. Stor. nat. Genova* 15:327;
Roewer, 1942, *Katalog der Araneae*, 1:401 (in part); Bonnet, 1956, *Bibliographia Araneorum*
2(2): 1661; Levy & Amitai, 1981b, *Zool. J. Linn. Soc.* 72:48.

Length of male 3.4–5.0 mm, female 2.8–5.5 mm. Coloration of carapace greyish
brown (Figs 37–39). Sternum blackish (Fig. 42). Legs brown. Opisthosoma shining
black to grey, folium usually extends dorsally over sides; margins of folium in front
and on sides conspicuous on mottled, greyish-white background (Fig. 44); venter
greyish, mottled with dark, short stripes.

Male Chelicera: Proximal and distal teeth pointing away from axis of basal segment of
chelicera (pr, d, Fig. 45). Proximal tooth with small, accesory denticles at base and
higher on tooth (Fig. 46), or with only one denticle at either position (Fig. 45).

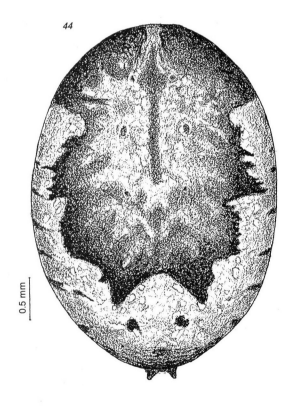

44

0.5 mm

Fig. 44: *Enoplognatha mandibularis* (Lucas, 1846); female,
opisthosoma, dorsal surface

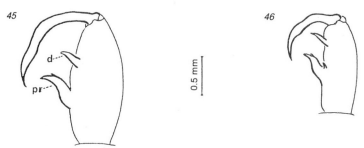

Figs 45–46: *Enoplognatha mandibularis* (Lucas, 1846); male, left chelicera
45. outer view; d – distal tooth, pr – proximal tooth; 46. outer view, variation

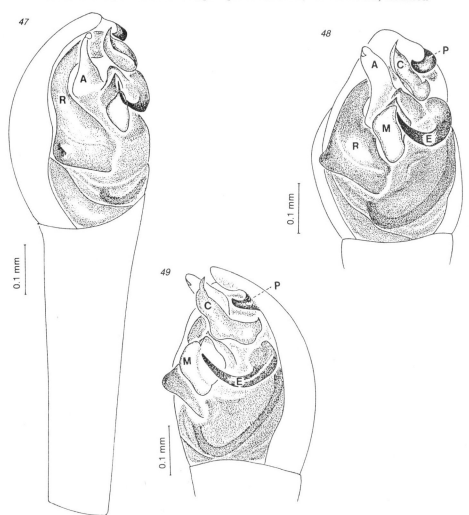

Figs 47–49: *Enoplognatha mandibularis* (Lucas, 1846); male, left palpus
47. mesal view; 48. ventral view; 49. retrolateral view; A – accessory apophysis,
C – conductor, E – embolus, M – median apophysis, P – paracymbial ‑hook, R – radix

Male Palpus: Tibia slender and appreciably longer than bulb (Fig. 47). Median apophysis (M) relatively small, stout and partly angular (Figs 47, 48). Accessory apophysis (A) large and slightly bent at apex (Figs 47, 48). Embolus (E) short, thick and only moderately arched (Fig. 49).

Female Epigynum: Dark with sclerotized, usually semicircular structure at centre; orifices not discernible (Fig. 50); posterior edge of epigynal plate extending as a lip over epigastric furrow (Fig. 50). Thick sclerotic ducts of spermathecae wind outwards, then converge and fuse on inner side with a thin brown sclerotic median plate (Fig. 51).

Distribution: Southern Europe, North Africa to East Mediterranean countries.

Israel: In the mountainous region from the Upper Galilee (1) to the Judean Hills (11) and along the Coastal Plain (4, 8).

Adult males are found from January to March, and mature females from January to May. Females with egg sacs fastened to the underside of stones were taken from March to May.

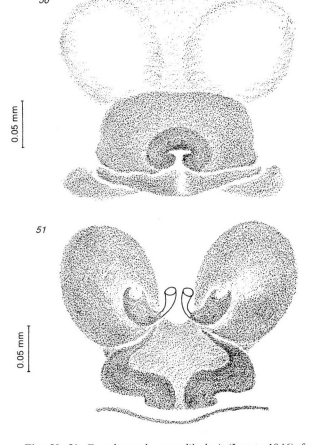

Figs 50–51: *Enoplognatha mandibularis* (Lucas, 1846); female
50. epigynum; 51. inner spermathecae, dorsal view

Enoplognatha macrochelis Levy & Amitai, 1981
Figs 40, 41, 43, 52–60

Enoplognatha macrochelis Levy & Amitai, 1981b, *Zool. J. Linn. Soc.*, 72:51.

Length of male 2.6–5.0 mm, female 4.0–5.7 mm. Coloration of prosoma yellowish brown with dark markings (Figs 40, 41, 52, 53). Legs light brown. Opisthosoma greyish with a distinct, dark dorsal folium bordered by broad, conspicuous margins; venter grey to black.

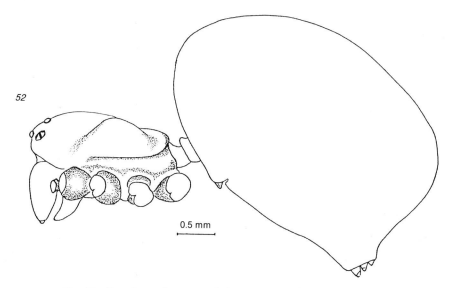

Fig. 52: *Enoplognatha macrochelis* Levy & Amitai, 1981; female, lateral view of spider (legs omitted)

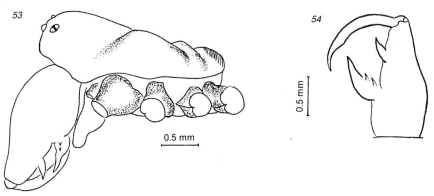

Figs 53–54: *Enoplognatha macrochelis* Levy & Amitai, 1981; male
53. prosoma, lateral view (legs omitted); 54. left chelicera, outer view

Male Chelicera: Large, with teeth pointing upwards and obliquely to axis of basal segment of chelicera (Fig. 54). Tip of proximal tooth extending to about height of distal tooth (Fig. 54); proximal tooth, at base, with two accessory denticles.

Male Palpus: Tibia shorter than bulb (Fig. 55). Median apophysis elongated and rounded on mesal side (Fig. 56); triangular, accessory apophysis relatively small, slender and pointed (Fig. 55); fine embolar tip strongly arched (Fig. 57).

Female Epigynum: Flattened orifice of central structure placed at posterior edge of epigynal plate (Fig. 58); dark, large swellings of internal organs visible above central structure. Round bodies of spermathecae relatively large; ducts form tight loop close to each spermathecal body, then converge into a thick, common duct (Fig. 59; variation, Fig. 60).

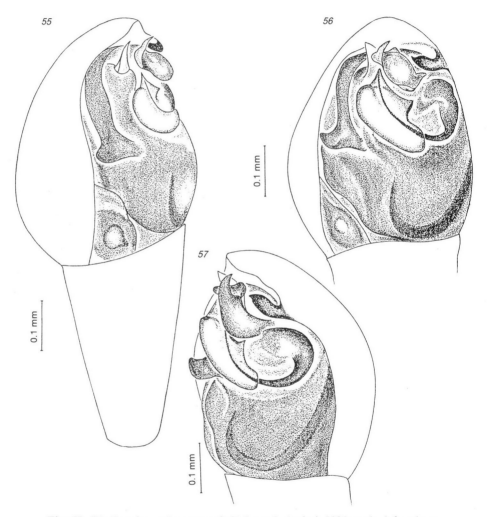

Figs 55–57: *Enoplognatha macrochelis* Levy & Amitai, 1981; male, left palpus
55. mesal view; 56. ventral view; 57. retrolateral view

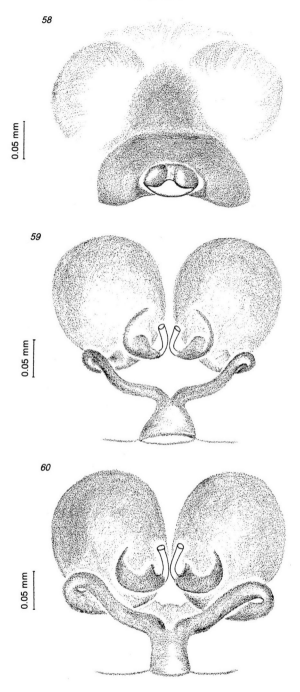

Figs 58–60: *Enoplognatha macrochelis* Levy & Amitai, 1981; female
58. epigynum; 59. inner spermathecae, dorsal view;
60. inner spermathecae, dorsal view, variation

Distribution: Greece, Israel, and possibly Jordan and Syria.

Israel: From the Golan Heights (18) through the central parts of the country and along the River Jordan (7, 13) to the Negev (15, 17).

Adults of both sexes are found from December–January to April. Females with egg sacs were encountered in April.

Enoplognatha deserta Levy & Amitai, 1981
Figs 61–67

Enoplognatha deserta Levy & Amitai, 1981b, *Zool. J. Linn. Soc.*, 72:55.

Length of male 5.7 mm, female 3.3–5.6 mm. Coloration of prosoma brown with dark margins (Fig. 61). Legs brown to yellowish-brown. Opisthosoma grey with a conspicuous folium on back; venter grey to black.

Male Chelicera: Large, with tip of proximal tooth not attaining height of distal tooth (Fig. 62). Distal tooth with two distinct accessory denticles at base, and proximal tooth with an indistinct accessory denticle at about two-thirds of its height (Fig. 62).

Male Palpus: Tibia about as long as bulb (Fig. 63). Bulb large; median apophysis relatively small, rounded on mesal side (Fig. 64); triangular accessory apophysis very large (Fig. 63); radix (R) with rounded bulges along mesal edges (Figs 64, 65); embolus thick and strongly arched (Fig. 65).

Female Epigynum: Opening of central structure partly divided by raised bottom (Fig. 66); area between opening and posterior convex rim on epigastric furrow, dark and appreciably raised (Fig. 66). Small, rounded, underlying spermathecal bodies usually discernible through integument. Ducts of spermathecae fused completely with sides of brown, sclerotic, median plate, widening thereafter into a broad, common duct (Fig. 67).

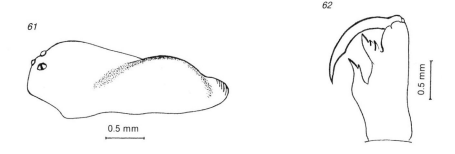

Figs 61–62: *Enoplognatha deserta* Levy & Amitai, 1981; male
61. prosoma, lateral view; 62. left chelicera, outer view

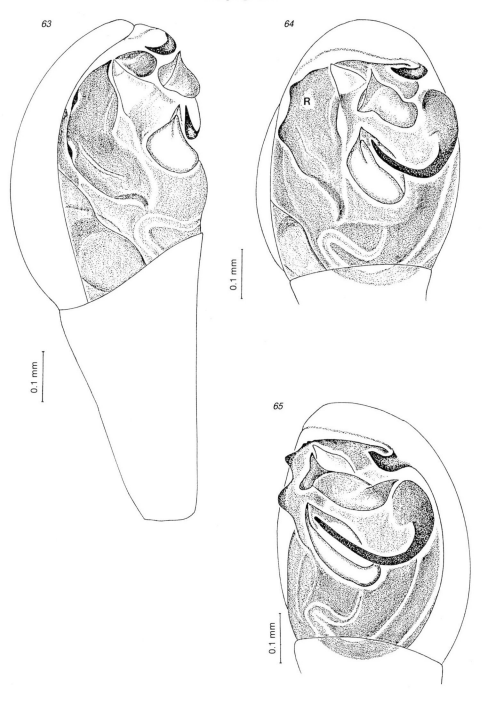

Figs 63–65: *Enoplognatha deserta* Levy & Amitai, 1981; male, left palpus
63. mesal view; 64. ventral view; R – radix; 65. retrolateral view

39

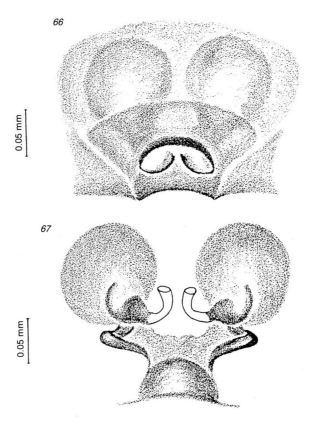

Figs 66–67: *Enoplognatha deserta* Levy & Amitai, 1981; female
66. epigynum; 67. inner spermathecae, dorsal view

Distribution: Israel, Egypt including Sinai to Tunisia, Algeria and Morocco.
Israel: Along the Dead Sea area (13) to the Negev (15, 17).
Egypt (Sinai): Wadi Ara'am (22).
An adult male was found in January, and mature females from January to April.

Enoplognatha parathoracica Levy & Amitai, 1981
Figs 68–75

Enoplognatha parathoracica Levy & Amitai, 1981b, *Zool. J. Linn. Soc.* 72:58

Length of male 2.9–4.0 mm, female 3.3–4.7 mm. Coloration of carapace brown to
deep brown with dark margins (Fig. 68). Sternum black, mottled with brown, and with
dark margins. Legs brown to yellowish-brown. Opisthosoma grey to black with
mid-dorsal and marginal rows of white spots (Fig. 69); dorsal median row consists

usually of two anterior pairs and two successive, single spots, the posterior spot placed above spinnerets; marginal rows each consist of a series of three single spots; sides and venter of opisthosoma grey mottled with white.

Male Chelicera: Apical part of basal segment of chelicera very wide and with raised swelling on outer side (Fig. 70). Strong, pointed proximal tooth armed with two accessory denticles; distal tooth only in form of a stout, stub-like denticle. (Fig. 70).

Male Palpus: Tibia appreciably shorter than bulb (Fig. 71). Median apophysis elongated and deeply notched apically (Figs. 71, 72); small accessory apophysis club-shaped, partly hidden beyond conductor (Fig. 71); embolus thick and strongly arched (Fig. 73).

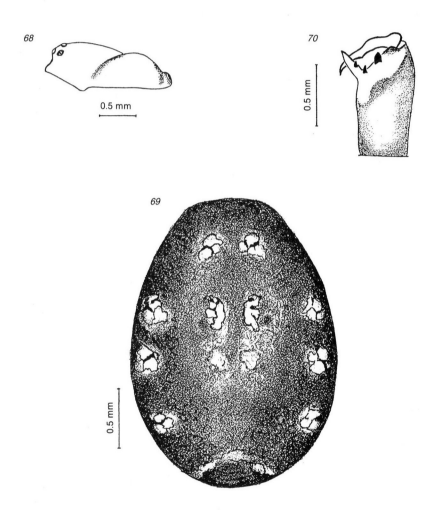

Figs 68–70: *Enoplognatha parathoracica* Levy & Amitai, 1981
68. male, prosoma, lateral view; 69. female, opisthosoma, dorsal surface;
70. male, left chelicera, outer view

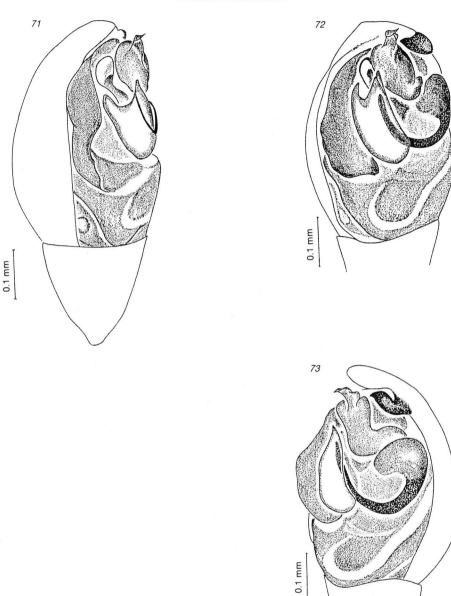

Figs 71–73: *Enoplognatha parathoracica* Levy & Amitai, 1981; male, left palpus
71. mesal view; 72. ventral view; 73. retrolateral view

Female Epigynum: Black, heavily sclerotized, swollen epigynal plate with two large, almost round, very distinct orifices (Fig. 74); raised space between orifices intact, clearly not traversed by seam or fissure. Thick, black, partly converging ducts of spermathecae become wider close to outer orifices (Fig. 75); dark, sclerotic, elongated plate at posterior edge of epigynal plate extending inwards across spermathecal orifices (Fig. 75).

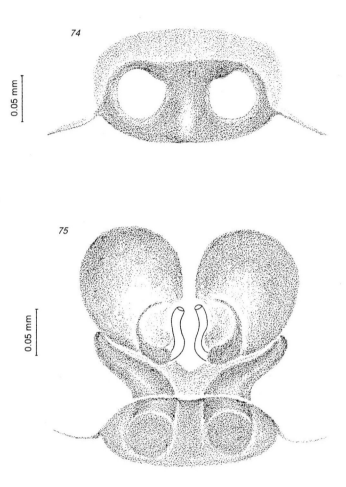

Figs 74–75: *Enoplognatha parathoracica* Levy & Amitai, 1981; female
74. epigynum; 75. inner spermathecae, dorsal view

Distribution: Israel, possibly in Lebanon.
Israel: From the Upper Galilee (1) to the Judean Hills (11) and northern Negev (15). Adult males are found from February to April, and mature females from March to May. Females with egg sacs were encountered in May.

Enoplognatha mediterranea Levy & Amitai, 1981
Figs 76–85

Enoplognatha mediterranea Levy & Amitai, 1981b, *Zool. J. Linn. Soc.* 72:62.

Length of male 3.6 mm, female 3.4–5.0 mm. Coloration of carapace deep to light brown with dark margins (Fig. 76). Sternum shining brown mottled with small greyish dots. Legs brown with dark tips. Opisthosoma black to grey with two mid-dorsal pairs of white spots (Fig. 77); spots occasionally indistinct; sides greyish mottled with brown dots; venter black, also around spinnerets.

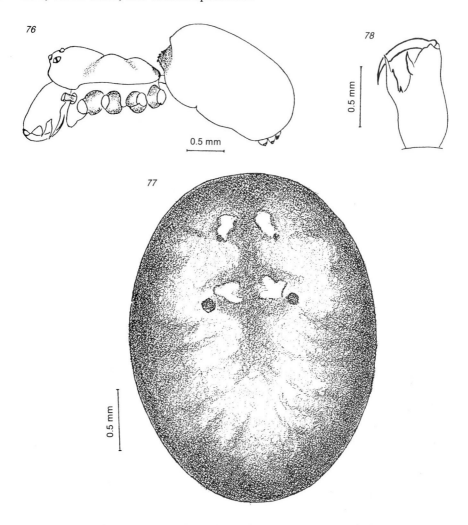

Figs 76–78: *Enoplognatha mediterranea* Levy & Amitai, 1981
76. male, lateral view of spider (legs omitted); 77. female, opisthosoma, dorsal surface;
78. male, left chelicera, outer view

Male Chelicera: Relatively small. Long, pointed proximal tooth attaining approximately height of distal tooth (Fig. 78); proximal tooth armed with two, small accessory denticles, one at base, another at about middle of height of tooth (Fig. 78).

Male Palpus: Relatively small. Tibia about as long as bulb (Fig. 79). Oval median apophysis apically notched (Fig. 79); thick, short, slightly triangular accessory apophysis ends with slightly bent tip (Fig. 80); lobe-like swelling projects from lower, mesal side of radix (Figs 79, 80); embolus long, fine and strongly arched (Fig. 81).

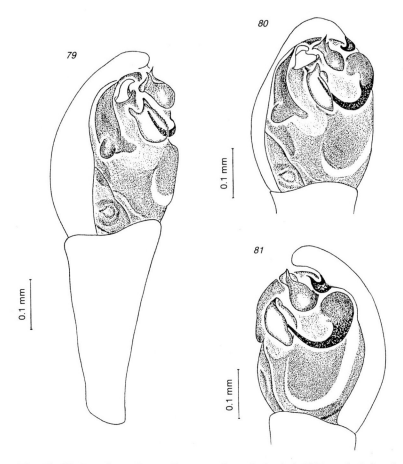

Figs 79–81: *Enoplognatha mediterranea* Levy & Amitai, 1981; male, left palpus
79. mesal view; 80. ventral view; 81. retrolateral view

Female Epigynum: Traversed by dark, bridge-like structure (Fig. 82; variation, Fig. 84); cup-like, sclerotic folds extend from inside bridge to posterior, slightly raised rim of epigynal plate (Figs. 82, 84). Relatively fine ducts of spermathecae bend medially towards each other, then diverge slightly close to outer orifices; brown, slightly arched, tube-like structure combines medially the two spermathecal bodies (Figs 83, 85).

45

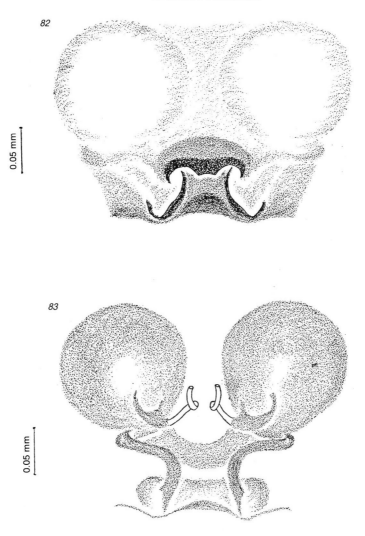

Figs 82–83: *Enoplognatha mediterranea* Levy & Amitai, 1981; female
82. epigynum; 83. inner spermathecae, dorsal view

Distribution: Israel, and probably Lebanon and Syria.
Israel: Mt Hermon (19) and Upper Galilee (1), along the Dead Sea (13) and Central Negev (17).
Adults were collected in February and March, and on Mt Hermon in April. Females were found under stones with egg sacs containing numerous eggs.

46

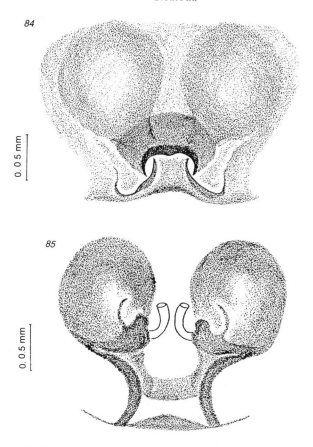

Figs 84–85: *Enoplognatha mediterranea* Levy & Amitai, 1981; female
84. epigynum, variation; 85. inner spermathecae, dorsal view, variation

Genus STEATODA Sundevall, 1833
Conspectus Arachnidum, Londini Gothorum, p. 16
Fig. 86

Type-species: *Araneus castaneus* Clerck, 1757.

Medium-sized to large theridiids, 2 to 13 mm total body length. Carapace longer than wide, in males sometimes rugose, and with a distinct median indentation (=fovea). Male with stridulatory ridges on posterior part of carapace. Size of eyes variable, lateral eyes touching each other or separated by less than their diameter. Chelicerae armed with one or two promarginal teeth, no retromarginal teeth in female; chelicerae of male often enlarged. First or fourth pair of legs longest, third shortest. Opisthosoma nearly oval, longer than wide or high, male with sclerotized ridges in front, above pedicel. Colulus very large.

47

86

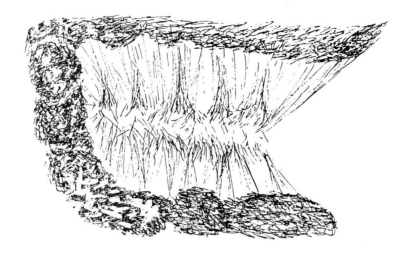

Fig. 86: *Steatoda*; web in a rock and earth cavity (diagrammatic)

Steatoda species, commonly known as cobweb spiders, construct a sparse, sheetlike web supported by partly viscous, sticky threads running in all directions (Fig. 86). The web may sometimes appear to consist of many small tents because of strongly tightened vertical threads. The spider builds no retreat and conceals itself during daytime wedged in a crevice. Species of *Steatoda* are found in rock clefts, on the ground, under stones and inside human habitations suspending their webs everywhere. In many species the opisthosoma is purplish-black with a white anterior belt; sometimes there is an additional dorsal pattern of lines or spots. *Steatoda* spiders feed on crawling and flying insects as well as on isopods (Crustacea); ants form part of the food of some species. *Steatoda* is distributed world-wide. About 100 species have been described, eleven are known from Israel and more are expected primarily from the southern parts.

Key to the species of **Steatoda** in Israel

Males:

1. Prosoma elongated, with posterior prolongation curving down sides as two sclerotized lobes behind tip of sternum (Fig. 152); small spiders, body less than 3 mm long
   S. erigoniformis (O.P.-Cambridge)
- Prosoma not modified; larger spiders   2
2. Endites of palpi armed with a few large protuberances (Fig. 87)
   S. albomaculata (DeGeer)
- Endites of palpi smooth or surface covered with fine granulations   3
3. Chelicerae relatively small with a slender fang (Figs 105, 136)   4
- Chelicerae large, robust with a thick fang (Figs 117, 125)   6

48

4. Prosoma and proximal segments of legs tuberculated 5
- Prosoma and legs smooth **S. latifasciata** (Simon)
5. Palpus with tibial segment about as long as bulb; paracymbial-hook (P) present on outside margin of cymbium (Figs 106, 108) **S. triangulosa** (Walckenaer)
- Palpus with tibial segment appreciably shorter than bulb; paracymbial-hook present inside back of cymbium **S. trianguloides** Levy
6. Dorsal surface of opisthosoma greyish white encircled with black, band-like patches (Fig. 123) **S. maura** (Simon)
- Opisthosoma on dorsum black with or without a pattern of white markings 7
7. Palpus elongated with tibia about as long as bulb; paracymbial-hook (P) present on outside margin of cymbium (Figs 118, 120) **S. paykulliana** (Walckenaer)
- Palpus with tibial segment appreciably shorter than bulb; paracymbial-hook present inside back of cymbium 8
8. Palpus with coiled embolus almost recurving on itself and a relatively slender median apophysis (Fig. 147) **S. grossa** (C.L. Koch)
- Palpus with terminal portion of embolus slightly arched at most and median apophysis broad with a sideways pointed extension (Fig. 100) **S. dahli** (Nosek)

Females:

1. Opisthosoma dorsally white with two rows of black, conspicuous spots (Fig. 131); epigynum narrow, straight at upper-anterior border (Fig. 132) **S. ephippiata** (Thorell)
- Opisthosoma black on dorsum, with or without a pattern of white spots or large markings; epigynum otherwise 2
2. Opisthosoma on back with distinctly separated pairs of small conspicuous, white spots, and a row of posterior spots, above spinnerets, but not with a single median row of spots or markings along dorsum (Figs 96, 97, 153) 3
- Opisthosoma dorsally with a white, median row of spots or light, continuous median markings or, dorsum evenly dark without spots 4
3. Opisthosoma, in front, with a fine, narrow white belt, sometimes discontinued at middle; venter with white, median spots or fine median line. Medium sized spiders

S. dahli (Nosek)
- Opisthosoma without anterior white belt or spots on venter. Small spiders, body about 3 mm long **S. erigoniformis** (O.P.-Cambridge)
4. Opisthosoma on ventral side, behind epigastric furrow, with a distinct, median line or, a white three-pronged fork 5
- Venter of opisthosoma otherwise marked or without markings 6
5. Venter of opisthosoma with a single, white, median line, and dorsally with a conspicuous light, broad median marking (Fig. 134); epigynum as in Fig. 140 **S. latifasciata** (Simon)
- Venter of opisthosoma with a white, three-pronged fork marking (Fig. 89); dorsum with large, partly contiguous, median spots (Fig. 88); epigynum as in Fig. 94

S. albomaculata (DeGeer)
6. Opisthosoma on ventral side, in front of spinnerets, with two white spots, very distinct on the dark background or, with two dark, elongated bars, visible on light background 7
- Venter of opisthosoma with only one indistinct spot behind epigastric furrow, or without light or dark markings 9

7. Venter of opisthosoma with two distinct white spots; dorsal surface with light triangular markings (Fig. 104); epigynum as in Figs 109, 110   **S. triangulosa** (Walckenaer)
–   Venter of opisthosoma with black, elongated bars; epigynum otherwise   8
8. Opisthosoma on dorsum evenly purplish-black, at most with faint remnants of dark, encircling band; epigynum with large, posteriorly directed, raised projection (Fig. 129)
          S. maura (Simon)
–   Dorsal surface of opisthosoma black with a restricted, central light marking, sometimes indistinct; rims of epigynum with no inwards projection (Fig. 142)
          S. xerophila Levy & Amitai
9. Opisthosoma usually purplish-black without any markings; epigynum divided by a raised median septum (Figs 149, 150)   **S. grossa** (C.L. Koch)
–   Opisthosoma usually with a large, coloured, median marking and a very broad light anterior belt (Fig. 116); posterior border of narrow epigynal opening bulging strongly anteriorly (Fig. 121)   **S. paykulliana** (Walckenaer)

Steatoda albomaculata (DeGeer,1778)

Figs 87–95

Aranea albomaculata De Geer, 1778, *Mémoires pour servir à l'histoire des Insectes* 7:257, pl. 15, figs 2–4.

Steatoda albomaculata —. Sundevall, 1833, *Conspectus Arachnidum*, Londini Gothorum, p. 17; Roewer, 1942, *Katalog der Araneae*, 1:405 (as *Lithyphantes*); Bonnet, 1957, *Bibliographia Araneorum* 2 (3):2550 (as *Lithyphantes*); Levi, 1957, *Bull. Mus. comp. Zool. Harv.* 117:396; Levy & Amitai, 1982b, *Zool. Scr.* 11:15.

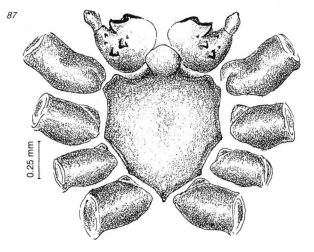

Fig. 87: *Steatoda albomaculata* (DeGeer, 1778); male, prosoma, ventral view, palpal endites with protuberances

50

Length of male 3.6–4.1 mm, female 5.2–5.9 mm. Coloration of prosoma brown to blackish, without granulation. Lateral eyes not touching. Endites of male palpi with distinct protuberances (Fig. 87). Legs yellowish to brown, in part, with dark articulations. Opisthosoma brown to black, almost encircled by light band; dorsum with a series of white median spots, sometimes merged into large blotches (Fig. 88); venter almost black with white, three-pronged fork-like marking (Fig. 89).

Male Chelicera: Basal segment, on frontal surface near tip with bulging, ridge-like fold; inner side armed with large coned tooth and bears fine bristles (Fig. 90).

Male Palpus: Tibia short. Black, thick, strongly angular, median apophysis pointing finger-like away from bulb (Figs 91–93); fine, filiform portion of embolus partly concealed at middle of bulb by membranous fold, and more apically, by large tapering conductor (Figs 92, 93).

Female Epigynum: Relatively large. Plate centrally traversed by broad, sclerotized, raised band widening on sides (Fig. 94); transverse band sometimes deeply notched at middle; shallow space between transverse band and epigastric furrow bordered on sides by light, fleshy folds (Fig. 94). Yellow, compact spermathecae connected to brown ducts, coiling underneath and widening as black tubes above spermathecae (Fig. 95).

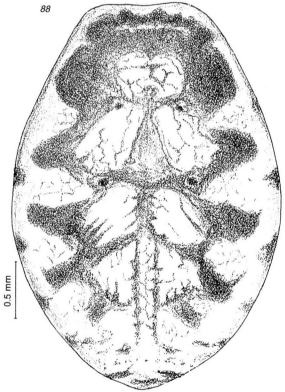

88

0.5 mm

Fig. 88: *Steatoda albomaculata* (DeGeer, 1778); female,
opisthosoma, dorsal surface

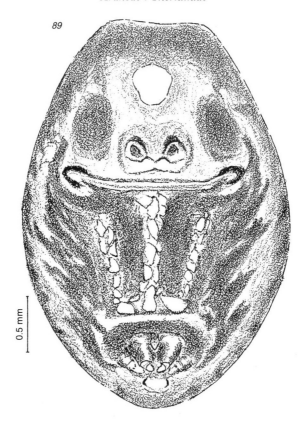

Fig. 89: *Steatoda albomaculata* (DeGeer, 1778); female,
opisthosoma, ventral surface

Fig. 90: *Steatoda albomaculata* (DeGeer, 1778); male,
tip of left chelicera, inner view

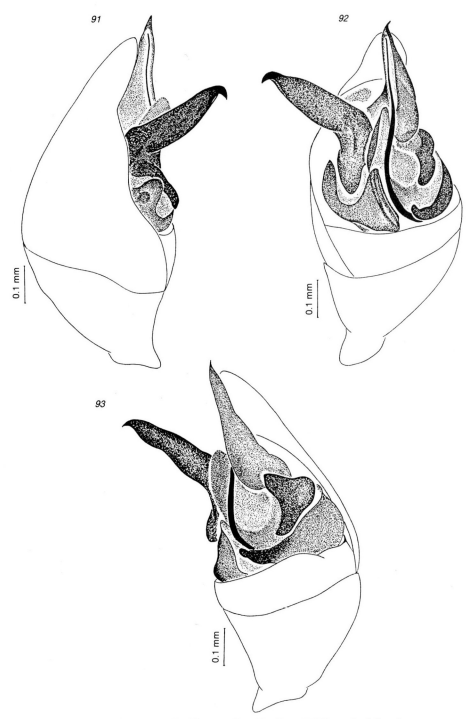

Figs 91–93: *Steatoda albomaculata* (DeGeer, 1778); male, left palpus
91. mesal view; 92. ventral view; 93. retrolateral view

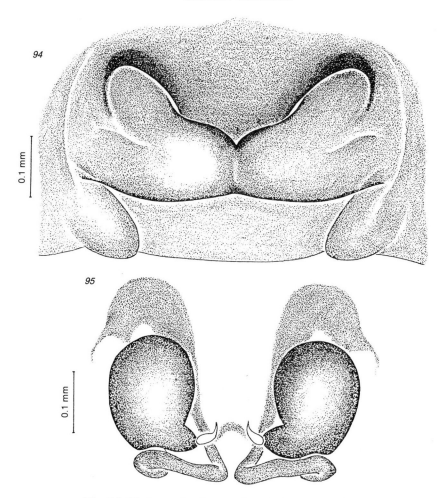

Figs 94–95: *Steatoda albomaculata* (DeGeer, 1778); female
94. epigynum; 95. inner spermathecae, dorsal view

Distribution: North and South America, northern Africa, Euroasia.
Israel: In the mountainous region from the Upper Galilee (1) and the Golan Heights
(18) to Jerusalem and Judean Hills (11), and in the northern Coastal Plain (4).
Adults are found under stones from April to July. Egg sacs are suspended in the web
and are coated with earth particles. Ant predators.

Steatoda dahli (Nosek, 1905)
Figs 96–103

Lithyphantes dahli Nosek, 1905, *Annln naturh. Mus. Wien*, 20:130, pl. 4, fig. 11; Roewer, 1942,
    Katalog der Araneae, 1:404; Bonnet, 1957, *Bibliographia Araneorum* 2 (3):2555.
Steatoda dahli —. Levy & Amitai, 1982b, *Zool. Scr.* 11:15.

Length of male 4.3 mm, female 5.0–6.3 mm. Coloration of carapace brown, in male with granulation, mainly on sides. Sternum brown encircled with black. Lateral eyes almost touching. Legs brown with dark distal segments. Opisthosoma black with short, fine, white belt in front, sometimes discontinued at middle; dorsum with variable number of bright, white spots: from one pair of spots on upper sides to two and three pairs in middle along with one or two pairs on sides, and additional, partly contiguous, single spot placed above spinnerets (Figs 96, 97); venter with one white spot behind epigastric furrow and a slightly widened marking in front of spinnerets, markings sometimes partly connected by faint white median line.

Male Chelicera: Large, with pointed tooth placed on mesal side (Fig. 98); fang slightly arched with fine indentations on inner side.

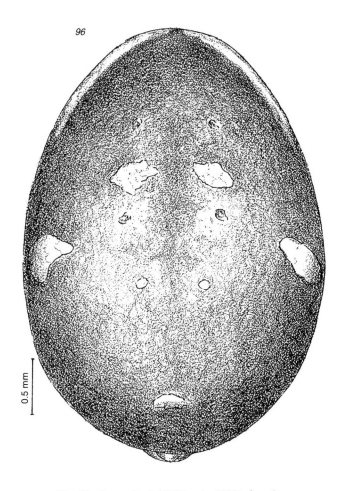

Fig. 96: *Steatoda dahli* (Nosek, 1905); female,
opisthosoma, dorsal surface

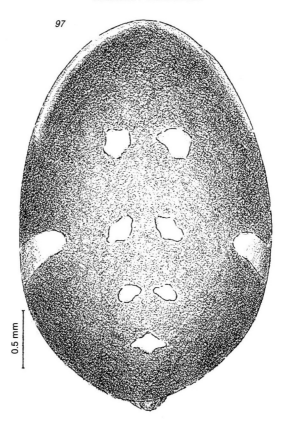

Fig. 97: *Steatoda dahli* (Nosek, 1905); female,
opisthosoma, dorsal view, variation

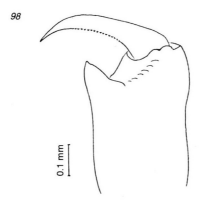

Fig. 98: *Steatoda dahli* (Nosek, 1905); male,
tip of left chelicera, inner view

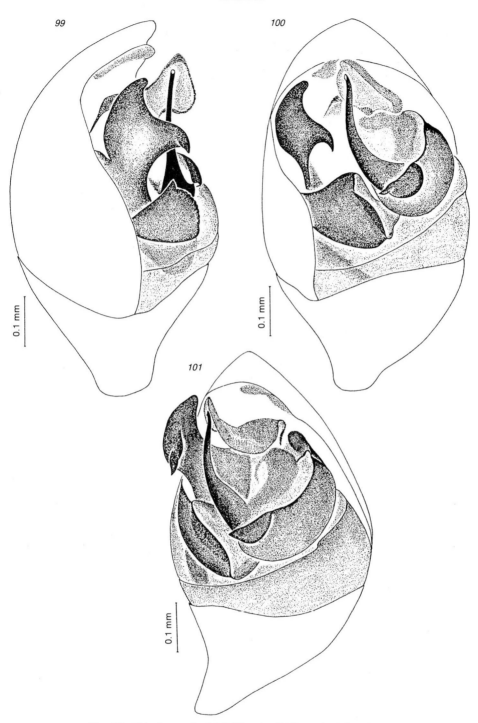

Figs 99–101: *Steatoda dahli* (Nosek, 1905); male, left palpus
99. mesal view; 100. ventral view; 101. retrolateral view

Male Palpus: Tibia short. Large and broad median apophysis inwardly curved, bears two apical extensions: one pointing distally, the other, slightly hooked, bending downwards (Figs 99–101); dark, small peculiar fold projecting at base of thick, arched embolus (Figs 100, 101); broad conductor extending only above tip of embolus (Figs 99–101).

Female Epigynum: Relatively large. Plate centrally traversed by yellow, bridge-like structure (Fig. 102); dark internal organs partly visible through brown, transparent integument (Fig. 102). Blackish, compact spermathecae directly appressed to thick, brown tubes (Fig. 103).

Distribution: South Turkey, Israel, presumably in Syria and Lebanon.

Israel: Mt Hermon (2025 m; 19) and Golan Heights (1075 m; 18).

Adults were collected in June and July. All were found under stones at altitude of over 1000 m, the Turkish holotype included.

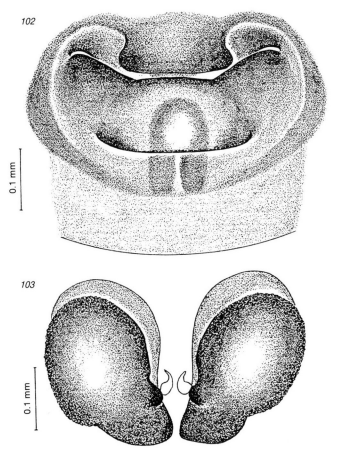

Figs 102–103: *Steatoda dahli* (Nosek, 1905); female
102. epigynum; 103. inner spermathecae, dorsal view

Steatoda triangulosa (Walckenaer, 1802)
Figs 104–111

Aranea triangulosa Walckenaer, 1802, *Fauna parisienne* 2:207.

Steatoda triangulosa —. Thorell, 1873, *Remarks on synonyms of European spiders* 4:505;
   Roewer, 1942, *Katalog der Araneae*, 1:416 (as *Teutana*); Levi, 1957, *Bull. Mus. comp. Zool.*
   Harv. 117:407; Bonnet, 1959, *Bibliographia Araneorum* 2(5):4377 (as *Teutana*); Levy &
   Amitai, 1982b, *Zool. Scr.* 11:17.

Length of male 2.5–4.7 mm, female 3.4–7.5 mm. Coloration of prosoma light to deep
brown, in male surface densely covered with small, pointed tubercles bearing bristles
(Fig. 104). Lateral eyes touching. Legs yellowish brown with dark patches; proximal

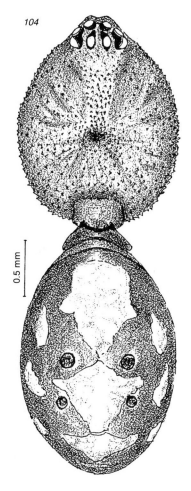

Fig. 104: *Steatoda triangulosa* (Walckenaer, 1802); male,
dorsal view of spider

segments in male, tuberculated. Opisthosoma shiny brown to purplish black with white spots; occasional specimens yellowish to light brown; white, large triangular or rhomboidal median spots on dorsum, sometimes merging with each other along middle portion, or spreading considerably down sides (Fig. 104); opisthosoma on sides usually with a row of about three white spots; venter with only two distinct bright spots in front of spinnerets; in male, on venter, brown anterior portion distinctly raised.

Male Chelicera: Relatively slender, with small, almost indistinct mesal denticle (Fig. 105).

Male Palpus: Slender with bulb about as long as tibia (Fig. 106). Median apophysis with widened, rounded base and S-shaped, apical portion (Figs 106, 107); filiform portion of embolus very short, hardly attaining height of median apophysis (Figs 106–108); conductor slender; paracymbial-hook (P) on retrolateral margin of cymbium, sometimes distinct (Fig. 108).

Female Epigynum: Often covered with brown, hardened secretion. Dark, swollen central structure traversed by narrow opening, slightly widening on sides; wall above opening sometimes with deep central concavity. Form of transverse opening varies slightly (Figs 109, 110); different impressions from the same epigynum may result from view taken at different angles. Dark brown, slightly oval, compact spermathecae, obliquely placed with only short, partly concealed, thick tubes (Fig. 111).

Distribution: Holarctic, but may have spread farther through human introductions. Israel: Very common throughout the country in the field as well as inside human habitations.

Adults of both sexes are found throughout the year, females with egg sacs, mainly from March to August. Specimens of all stages are often found living under the same log, in close quarters. Egg sacs, up to 17, are suspended in the web with the female. Size of an egg sac is about 5 mm, it contains about 50 eggs. Hatching occurs after a month or a little more, often after oviposition, and the young go through three to five moults, attaining maturity after three to four months.

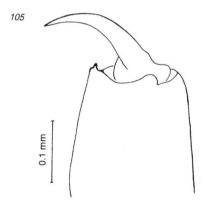

Fig. 105: *Steatoda triangulosa* (Walckenaer, 1802); male, tip of left chelicera, inner view

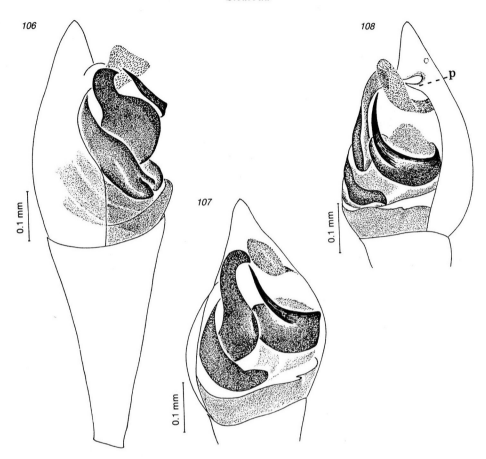

Figs 106–108: *Steatoda triangulosa* (Walckenaer, 1802); male, left palpus
106. mesal view; 107. ventral view; 108. retrolateral view; P – paracymbial-hook

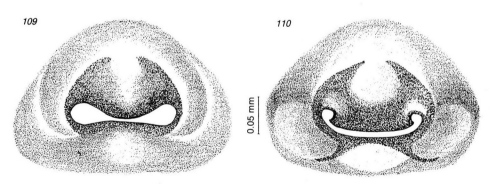

Figs 109–110: *Steatoda triangulosa* (Walckenaer, 1802); female
109. epigynum; 110. epigynum, variation

61

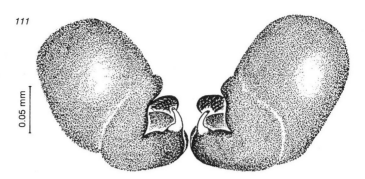

Fig. 111: *Steatoda triangulosa* (Walckenaer, 1802); female,
inner spermathecae, dorsal view

Steatoda trianguloides Levy, 1991
Figs 112–115

Steatoda trianguloides Levy, 1991, *Bull. Br. arachnol. Soc.* 8(7):228.

Length of male 3.3 mm; female unknown. Coloration of prosoma light brown; surface densely covered with small, pointed tubercles. Legs yellowish and tuberculated. Opisthosoma light with a few black spots on dorsum (Fig. 112).

Male Chelicera: Relatively small, without denticles.

Male Palpus: Slender with bulb appreciably longer than tibia (Fig. 113). Slender, cylindrical median apophysis rises almost straight apically (Fig. 113); thick, blackish process projects from basal embolar division (Figs 114, 115); thick winding embolus tapers apically to a straight filiform portion (Figs 114, 115). Paracymbial-hook present inside back of cymbium.

Distribution. Israel: Mt Hermon, 1700 m; July (19).

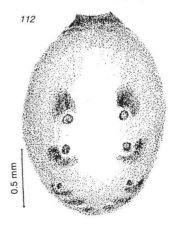

Fig. 112: *Steatoda trianguloides* Levy, 1991; male,
opisthosoma, dorsal surface

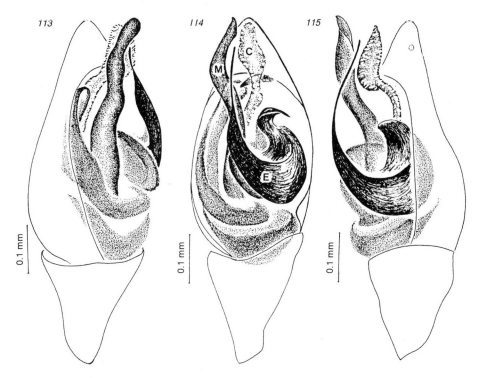

Figs 113–115: *Steatoda trianguloides* Levy, 1991; male, left palpus
113. mesal view; 114. ventral view; C – conductor, E – embolus,
M – median apophysis; 115. retrolateral view

Steatoda paykulliana (Walckenaer, 1806)
Figs 116–122

Theridion paykullianum Walckenaer, 1806, *Histoire naturelle des Aranéides*, pl. 4.
Steatoda paykulliana —. Simon, 1875, *Annls Soc. ent. Fr.* (5)5 *Bull.*:197; Roewer, 1942,
    Katalog der Araneae, 1:404 (as *Lithyphantes*); Bonnet, 1957, *Bibliographia Araneorum*
    2(3):2557 (as *Lithyphantes*); Levy & Amitai, 1982b, *Zool. Scr.* 11:18.

Length of male 4.9–8.4 mm, female 9.1–12.9 mm. Carapace and sternum deep brown
to black, without granulations (Fig. 116). Lateral eyes not touching. Legs light to dark
brown near articulations. Opisthosoma shiny black, encircled in front and partly on
sides by broad belt of yellow, orange or red; dorsum usually with yellow to red, wide
conspicuous, median marking with three to five branches (Fig. 116); anterior branch
may extend down sides to encircling belt, dividing the black space on dorsum into two
portions; sides of opisthosoma black with one or two coloured blotches behind
encircling belt; venter with one small, light spot close to epigastric furrow.

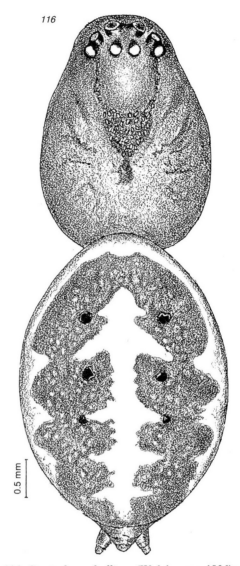

116

Fig. 116: *Steatoda paykulliana* (Walckenaer, 1806); male,
dorsal view of spider

Male Chelicera: Relatively large and robust. Apical margin armed with strong, broad tooth and thick sclerotized ridge with pointed tips (Fig. 117); fang large with distinctly indented base.

Male Palpus: Large and elongated. Concave, median apophysis relatively short, with tip hook-like inclined (Figs 118, 119); arched filiform portion of embolus extending above median apophysis; conductor with much widened apical portion (Figs 119, 120); retrolateral margin of cymbium with a distinct paracymbial-hook (P; Fig. 120).

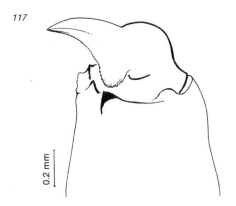

Fig. 117: *Steatoda paykulliana* (Walckenaer, 1806); male,
tip of left chelicera, inner view

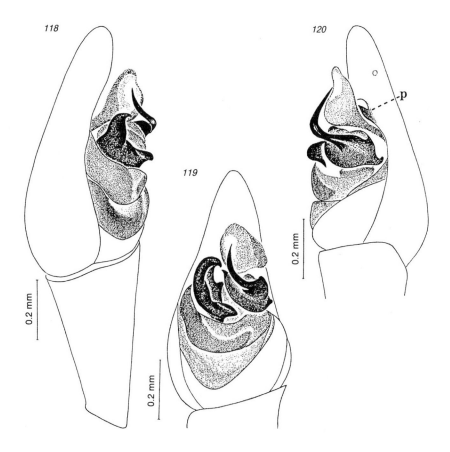

Figs 118–120: *Steatoda paykulliana* (Walckenaer, 1806); male, left palpus
118. mesal view; 119. ventral view; 120. retrolateral view; P – paracymbial-hook

Female Epigynum: Brown, large and markedly raised mainly on posterior portion (Fig. 121); plate traversed by wide opening, but at middle almost divided by strong forward projecting bulge of posterior edge (Fig. 121). Dark brown, very large, compact spermathecae with thick, short tubes, hardly visible on dorsal side (Fig. 122).

Distribution: Southern Europe to the Balkans and Russia, throughout northern Africa to Ethiopia, Middle East to Yemen.

Israel: Common throughout the country.

This is the largest *Steatoda* of this region. Adults of both sexes are found from autumn to early summer, November to April–May, females also to July. Females with one or two egg sacs are found mainly in April–May. Size of an egg sac may reach 15 mm or more containing several hundred eggs. Juveniles attain maturity after about eight to ten months.

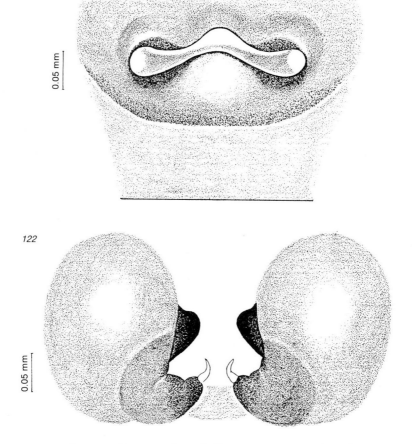

Figs 121–122: *Steatoda paykulliana* (Walckenaer, 1806); female
121. epigynum; 122. inner spermathecae, dorsal view

Steatoda maura (Simon, 1909)
Figs 123–130

Lithyphantes maurus Simon, 1909, *Mem. r. Soc. esp. Hist. nat.* 6:23.
Steatoda maura —. Levy & Amitai, 1982b, *Zool. Scr.* 11:21.

Length of male 6.0–8.8 mm, female 7.7–10.6 mm. Coloration of carapace and sternum light brown to brown, in male, with fine granulation mainly on sides and posteriorly. Lateral eyes almost touching. Legs yellow to brown with dark parts near articulations, in male also with small granules. Opisthosoma of adult male and young female white on dorsum, encircled by black, band-like shaped patches (Fig. 123); posterior blackish; venter light, occasionally with dark markings. Opisthosoma of adult female purplish-black on back surrounded anteriorly and on sides by white creamy, broad belt; venter usually light with two purplish-black, parallel bars placed longitudinally (Fig. 124). Sometimes, faint dorsal marking of young stage still slightly visible in adult.

123

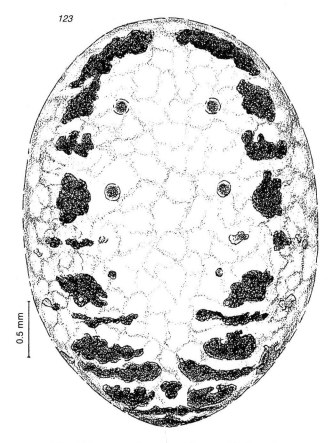

Fig. 123: *Steatoda maura* (Simon, 1909); male,
opisthosoma, dorsal surface

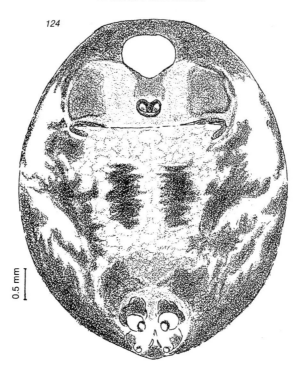

124

0.5 mm

Fig. 124: *Steatoda maura* (Simon, 1909); female,
opisthosoma, ventral surface

Male Chelicera: Large, with strong mesal tooth and thick fang with fine indentations on inner side (Fig. 125).

Male Palpus: Elongated, with long, slender, apically bent, median apophysis (Figs 126, 127); strong, winding embolus tapering distally almost attaining top of cymbium (Figs 126–128); conductor spindle-shaped, with concavity on mesal side (Figs 126, 127).

Female Epigynum: At centre with a broad, tongue-shaped swelling pointing towards epigastric furrow (Fig. 129); greyish, membranous area, partly surrounding central swelling, bordered on outside by brown, raised rim (Fig. 129); posterior, sclerotized rim of greyish membrane narrow and markedly raised. Brown to blackish spermathecae large, tubes not visible in dorsal view (Fig. 130).

Distribution: Morocco, Israel, Greece.

Israel: Eastern Samaria (6), Judean Desert (12), Dead Sea area (13) to Central Negev (17).

Adult males were found only from September to November, while adult females are found from October throughout the winter to May. Females with egg-sacs are usually found in March–April. Egg sacs are about 10–12 mm in size and contain several hundred eggs. The young hatch about six weeks after oviposition and reach maturity

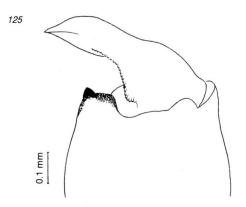

Fig. 125: *Steatoda maura* (Simon, 1909); male,
tip of left chelicera, inner view

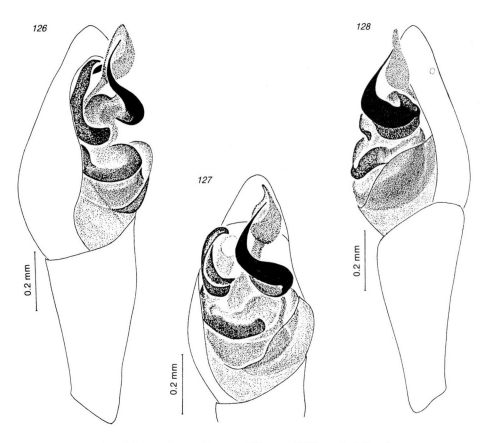

Figs 126–128: *Steatoda maura* (Simon, 1909); male, left palpus
126. mesal view; 127. ventral view; 128. retrolateral view

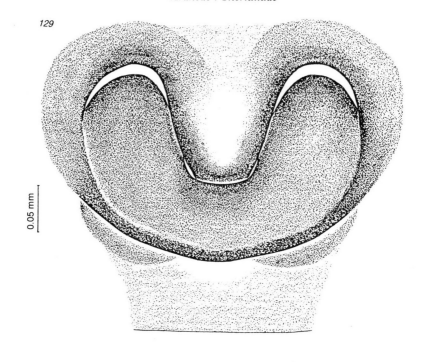

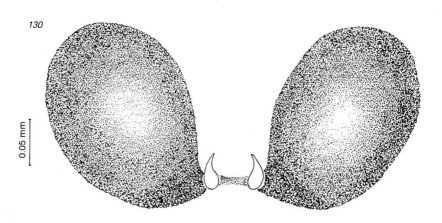

Figs 129–130: *Steatoda maura* (Simon, 1909); female
129. epigynum; 130. inner spermathecae, dorsal view

in five to seven months. Males moult about seven times and females a few times more before attaining maturity. Their web is usually spun under overhanging rock shelves and on low bushes, but not under stones.

Steatoda ephippiata (Thorell, 1875)

Figs 131–133

Lithyphantes ephippiatus Thorell, 1875, *K. svenska VetenskAkad. Handl.* (N.F.) 13(5):63.
Steatoda ephippiata —. Levy & Amitai, 1982b, *Zool. Scr.* 11:22.

Adult male unknown. Length of female 6.3–7.1 mm. Coloration of prosoma and legs brown. Lateral eyes not touching. Opisthosoma on dorsum usually white with a series of black spots arranged in two longitudinal rows (Fig. 131); anterior spots may merge to form one dark bar; ventral surface black with bright, white median line widening slightly posteriorly; median line extends backwards from behind epigastric furrow about two-thirds of distance to spinnerets.

Female Epigynum: Relatively narrow. Centre traversed by white, opaque membranous area, widening on sides and bordered laterally by brown, slender arched rims (Fig. 132); anterior-upper tips of arched lateral rims widely separated from each other, recurving under straight, anterior-upper edge of central, opaque area (Fig. 132); transparent bulge on basal, middle portion of transverse structure, protruding upwards, away from epigastric furrow (Fig. 132). Round spermathecae light brown in centre, darker on sides; dark, tight tubes, partly visible only behind spermathecae, on space near epigastric furrow (Fig. 133).

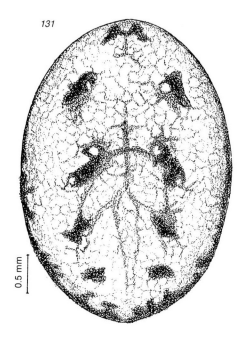

131

0.5 mm

Fig. 131: *Steatoda ephippiata* (Thorell, 1875); female,
opisthosoma, dorsal surface

71

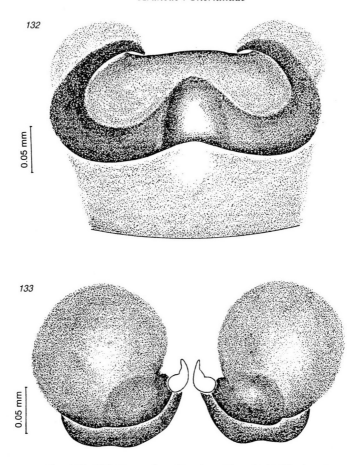

Figs 132–133: *Steatoda ephippiata* (Thorell, 1875); female
132. epigynum; 133. inner spermathecae, dorsal view

Distribution: Algeria, Tunisia, Libya, Egypt including Sinai, southern Israel.
Israel: Ze'elim (15), Makhtesh Ramon (16).
Egypt (Sinai): Bir Gifgafa (21), 'Ayun Mussa (23).
The few adult specimens collected thus far in the Middle East were found in March–April, and one in September

Steatoda latifasciata (Simon, 1873)
Figs 134–141

Lithyphantes latifasciatus Simon, 1873, *Mém. Soc. r. Sci.* Liège (2)5:83, pl. 2, fig. 31; Roewer, 1942, *Katalog der Araneae*, 1:404; Bonnet, 1957, *Bibliographia Araneorum* 2(3):2556.
Steatoda latifasciata —. Levy & Amitai, 1982b, *Zool. Scr.* 11:23.

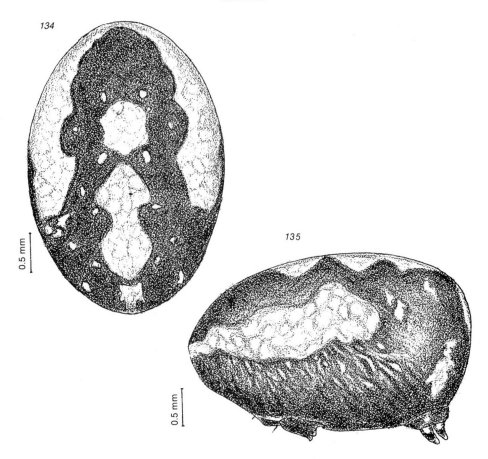

Figs 134–135: *Steatoda latifasciata* (Simon, 1873); female, opisthosoma
134. dorsal surface; 135. lateral view

Length of male 2.8–3.5 mm, female 3.8–6.1 mm. Carapace dark brown with black radiating lines. Sternum purplish-black. Lateral eyes usually not touching. Legs yellowish to deep brown. Opisthosoma purplish-black with white, very broad anterior belt, extending over upper sides of dorsum (Figs 134, 135); dorsum with conspicuous, white median marking, slightly constricted at a few points (Fig. 134); venter with white median line, widening slightly posteriorly, in front of spinnerets, and two additional, light spots, placed obliquely above spinnerets.

Male Chelicera: Relatively small with mesal, apical edge pointed and slightly raised (Fig. 136); fang slender.

Male Palpus: Tibia short. Slender, elongated median apophysis with partly pointed, strongly bent tip (Figs 137–139); arched, fine filiform portion of embolus rising high above median apophysis; conductor carried on long slender stalk, with widened terminal portion surrounding a slight concavity (Figs 137–139).

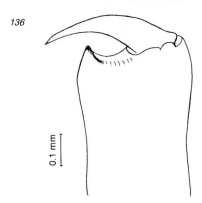

Fig. 136: *Steatoda latifasciata* (Simon, 1873); male,
tip of left chelicera, inner view

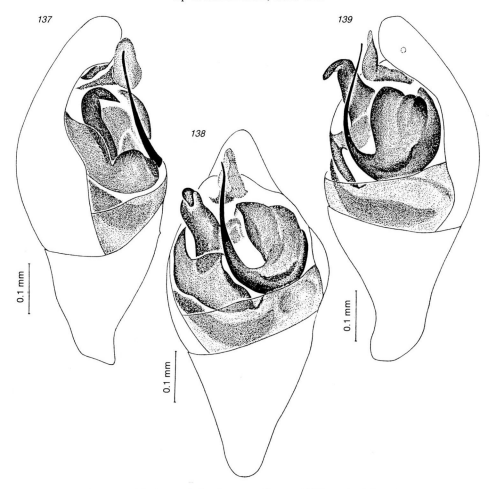

Figs 137–139: *Steatoda latifasciata* (Simon, 1873); male, left palpus
137. mesal view; 138. ventral view; 139. retrolateral view

74

Female Epigynum: White, opaque central space partly surrounded by brown, fine, raised rims (Fig. 140); small, deep depression placed in middle, above epigynal plate, bordered posteriorly by brown, raised short upper edge of central, opaque area (Fig. 140); sclerotized posterior margin of epigynum markedly raised above epigastric furrow. Purple black spermathecae with thick tubes only partly visible along median line (Fig. 141).

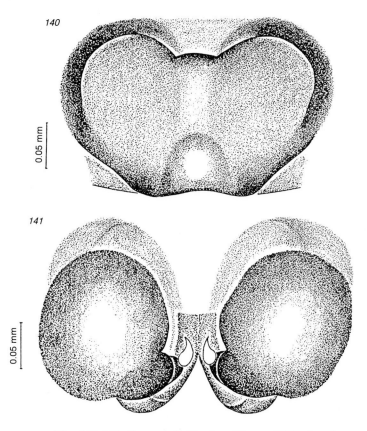

Figs 140–141: *Steatoda latifasciata* (Simon, 1873); female
140. epigynum; 141. inner spermathecae, dorsal view

Distribution: Morocco, Sicily, Israel, Egypt (Sinai).
Israel: Along the Jordan Valley (7), Dead Sea area (13) and Judean Desert (12) to the southern Coastal Plain (9), the Negev (15, 16) and the Arava Valley (22).
Egypt (Sinai): Near Mitla pass (21) and east of Sinai mountains (22).
Adult males and females are found almost throughout the year, except for August–September. They spin webs on the ground, some also in sand dunes, inside and around abandoned burrows and fissures. The spherical egg sac is about 5 mm in size, contains over 30 eggs, hardly discernible through the silken walls.

Steatoda xerophila Levy & Amitai, 1982
Figs 142, 143

Steatoda xerophila Levy & Amitai, 1982b, *Zool. Scr.* 11:24.

Male unknown. Length of female 6.3–6.6 mm. Coloration of prosoma brown. Lateral eyes almost touching. Legs yellowish-brown. Opisthosoma purplish-black with white, median spot or small blotch on back, and fine, narrow white belt anteriorly; venter light with two black bars on space behind epigastric furrow.

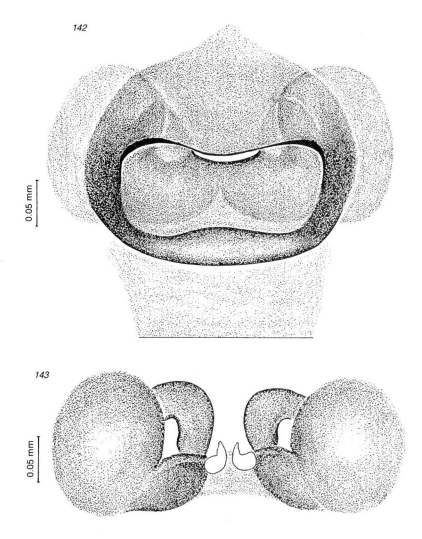

Figs 142–143: *Steatoda xerophila* Levy & Amitai, 1982; female
142. epigynum; 143. inner spermathecae, dorsal view

Female Epigynum: Centre traversed by large, wide depression encircled by thick, brown rims (Fig. 142); brown, posterior edge of epigynal plate, below central depression markedly raised above yellowish space extending to epigastric furrow (Fig. 142); parts of dark, internal organs visible through integument on bottom of central depression. Yellowish-brown spermathecae placed at a distance from each other, with brown thick, rounded extension projecting medially (Fig. 143); arching membranous tubes partly visible in background.

Distribution. Southern Israel: Negev (16, 17).

Adult females were collected in January and February.

Steatoda grossa (C.L. Koch, 1838)
Figs 144–151

Theridium grossum C.L. Koch, 1838, *Die Arachniden* 4:112, fig. 321.
Steatoda grossa —. C.L. Koch, 1851, *Uebersicht des Arachnidensystems* 5:17; Roewer, 1942,
  Katalog der Araneae 1:414 (as *Teutana*); Levi, 1957, *Bull. Mus. comp. Zool. Harv.* 117:404;
  Bonnet, 1959, *Bibliographia Araneorum* 2(5):4373 (as *Teutana*); Levy & Amitai, 1982b,
  Zool. Scr. 11:25.

Length of male 3.8–6.8 mm, female 5.9–10.3 mm. Coloration of carapace brown, of sternum yellowish-brown, male also with fine granules. Lateral eyes touching. Legs yellow to brown. Opisthosoma purplish-black with white markings on back (Fig. 144). Dorsal pattern consists usually of a broad, crescent-shaped mark in front, on upper side, a median row of partly rhomboidal spots, and two rows, one on each side, of creamy spots; anterior spots may merge and form one large blotch; ventral surface of opisthosoma usually light posteriorly, sometimes marked as large trapezoidal blotch; in male, anterior part dark and markedly swollen. Opisthosoma of old, adult female purplish-black, almost shiny, without light markings.

Male Chelicera: Small, stout tooth protruding from nearly straight upper edges; sometimes an additional denticle visible close to mesal tooth (Fig. 145); fang short, thick with fine indentations on inner side.

Male Palpus: Elongated with relatively short tibia (Figs 146–148). Brown, relatively small, median apophysis with tip medialwards inclined (Fig. 147); partly coiled embolus with fine, filiform portion extending high above median apophysis (Figs 146–148); membranous, elongated conductor attaining height of tip of cymbium (Figs 146–148).

Female Epigynum: Relatively large. Central depression often divided by very distinct, raised median septum (Figs 149, 150); central depression half encircled by raised rims, sometimes slightly projecting in middle into depression (Fig. 149); posterior borders of epigynal plate intact (Fig. 149) or partly notched at middle (Fig. 150); internal organs visible in part inside central depression. Blackish round spermathecae partly surrounded by large membranous fold (Fig. 151).

144

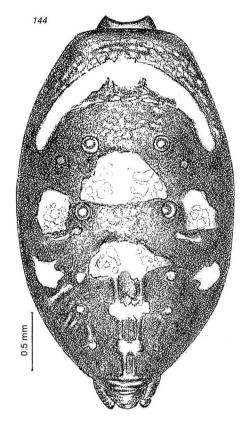

Fig. 144: *Steatoda grossa* (C.L. Koch, 1838); male,
opisthosoma, dorsal surface

145

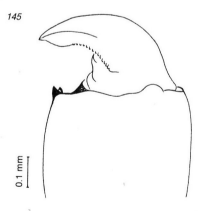

Fig. 145: *Steatoda grossa* (C.L. Koch, 1838); male,
tip of left chelicera, inner view

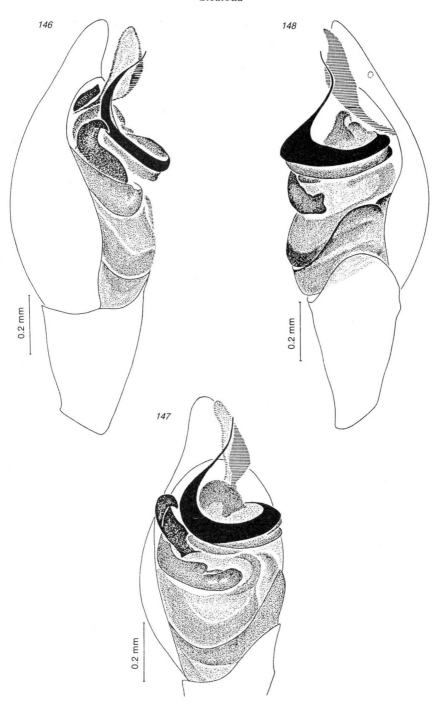

Figs 146–148: *Steatoda grossa* (C.L. Koch, 1838); male, left palpus
146. mesal view; 147. ventral view; 148. retrolateral view

Distribution: Cosmopolitan at least in association with human habitations.

Israel: Only inside houses, mainly in animal-breeding rooms kept under constant, high temperature, and an abundant supply of houseflies. Adults are found there throughout the year. A female may produce several egg sacs. The size of an egg sac is about 10 mm, and it contains over one hundred white eggs. The young hatch from the eggs after about one month; female may reach maturity in about six months. In laboratory, adult females have been kept alive for more than two years.

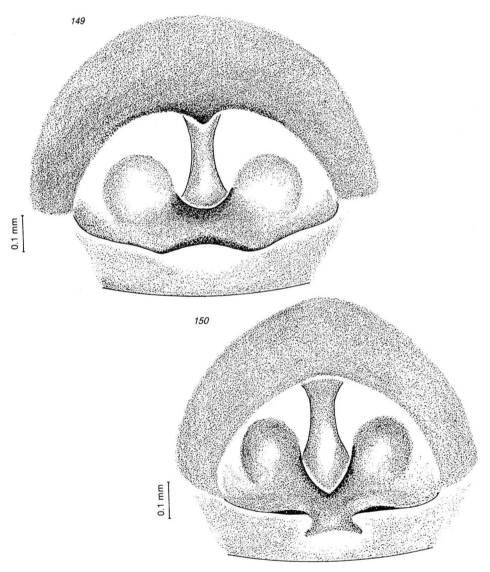

Figs 149–150: *Steatoda grossa* (C.L. Koch, 1838); female
149. epigynum; 150. epigynum, variation

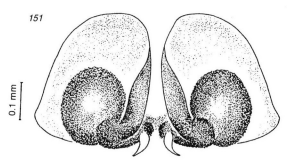

Fig. 151: *Steatoda grossa* (C.L. Koch, 1838); female,
inner spermathecae, dorsal view

Steatoda erigoniformis (O.P.-Cambridge, 1872)
Figs 152–160

Theridion erigoniforme O.P.-Cambridge, 1872, *Proc. zool. Soc. Lond.* 1872:284.
Steatoda erigoniformis —. Levi, 1962, *Psyche Camb.* 69:25; Levy & Amitai, 1982b, *Zool. Scr.*
   11:26.

Length of male 2.3–2.6 mm, female 2.4–3.1 mm. Carapace and sternum of male deep
brown with distinct granulation, in female, orange to light brown and smooth.
Carapace of male elongated with posterior prolongation curving to underside of
prosoma as two sclerotized lobes facing each other behind tip of sternum (Fig. 152).

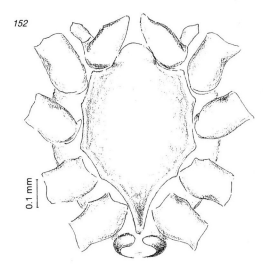

Fig. 152: *Steatoda erigoniformis* (O.P.-Cambridge, 1872); male,
prosoma, ventral view, extensions of carapace reach point behind tip of sternum

Lateral eyes touching. Legs dark brown to almost black on proximal parts, yellowish-orange on distal segments. Opisthosoma purplish-black with four spots on anterior upper sides of dorsum and two or three smaller ones in a row, above spinnerets (Fig. 153). Female with two additional white spots, obliquely placed, in front of spinnerets (Fig. 154).

Male Chelicera: A tooth is placed at base of a recess on inner side, and a swelling bearing small tubercles with bristles is found on the outer side (Fig. 155).

Male Palpus: Very small, with short tibial segments. Low median apophysis with hardly an apical extension (Fig. 156); embolus helix-like winding with filiform portion recurving on itself (Figs 157, 158); large membranous conductor partly surrounding tip of embolus (Figs 156–158).

Female Epigynum: Central depression with raised, partly transparent, bridge-like structure at middle (Fig. 159); upper corners of central structure partly covered by oblique, narrow, raised sclerotized arches (Fig. 159). Black spermathecae elongated, with tubes running medially and bending into funnelled, membranous folds (Fig. 160).

Distribution: World-wide, primarily cosmotropical.

Israel: Along the Mediterranean Coastal Plain (4, 9) and from the Golan Heights (18) through the Rift Valley (7, 13) to the Negev (15, 17).

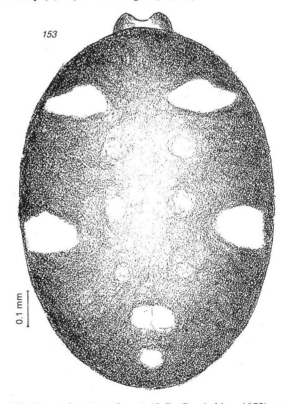

Fig. 153: *Steatoda erigoniformis* (O.P.-Cambridge, 1872); male, opisthosoma, dorsal surface

Considered the smallest *Steatoda* of this region. It lives under stones. Adult males were collected in June, August and December, and mature females in February, and from June to December.

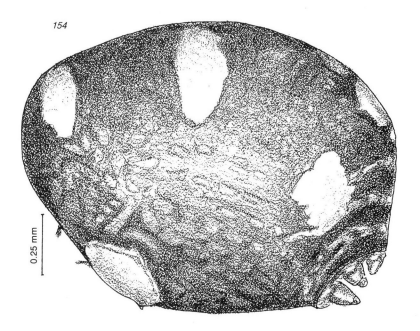

Fig. 154: *Steatoda erigoniformis* (O.P.-Cambridge, 1872); female, opisthosoma, lateral view

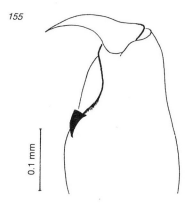

Fig. 155: *Steatoda erigoniformis* (O.P.-Cambridge, 1872); male, tip of left chelicera, inner view

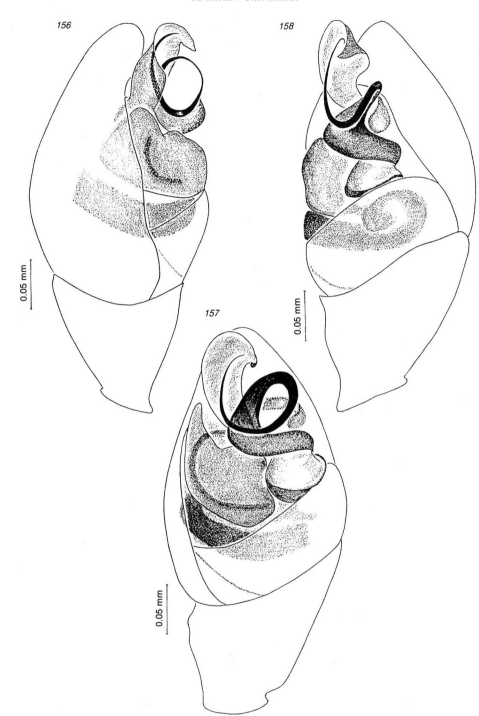

Figs 156–158: *Steatoda erigoniformis* (O.P.-Cambridge, 1872); male, left palpus
156. mesal view; 157. ventral view; 158. retrolateral view

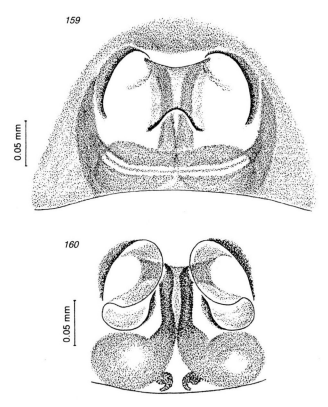

Figs 159–160: *Steatoda erigoniformis* (O.P.-Cambridge, 1872); female
159. epigynum; 160. inner spermathecae, dorsal view

Genus LATRODECTUS Walckenaer, 1805
Tableau des Aranéides, Paris, p. 81
Figs 161–164

Type-species: *Aranea tredecimguttata* Rossi, 1790.

Theridiids with remarkable sexual dimorphism: large female, some may attain body length of 20 mm, and much smaller males, occasionally five times smaller. Carapace longer than wide, sometimes about as long as wide. Lateral eyes widely separated (Fig. 161). Chelicerae without teeth (Fig. 162). Legs relatively thick and moderately long, first pair longer than fourth. Opisthosoma nearly globular (Fig. 163) displaying differences in spininess. Colulus large. Palpus of male with thick embolus, spirally coiled around modified cymbium. Epigynum of female traversed at middle by an elliptical opening; two internal seminal receptacles dumb-bell shaped, and connecting ducts, coiled.

Species of *Latrodectus*, commonly called Widows, are known for the lethal potency of the venom of certain species and their virulence in particular to mammals. When confronted by a large animal, the spider hides in the web and biting thus occurs when it is accidentally squeezed or pressed against the skin. The main signs and symptoms of *L. tredecimguttatus* bite are colicky abdominal pain with the abdomen tight and sensitive to touch, diffuse sweating, tremor, restlessness, extreme weakness and high blood pressure. Cure is attained usually by subcutaneous injection of atropine. Considering the spider's biology, the toxicity against arthropods appears to be of the most importance.

Latrodectus species are found throughout the world, mainly in the drier warmer regions. There is close similarity in the general morphology amongst all species and often also a considerable intraspecific variation of colour. Five sympatric species of *Latrodectus* are found in the more arid zones of Israel and one allopatric population of the cosmotropical *L. geometricus* occurs in the Mediterranean Coastal Plain. The elaborate palpal structure of the adult male of all eremic *Latrodectus* species of Israel, except that of *L. dahli*, is similar (Figs 172–174). The embolar coils may appear tightened or loose, and the basal embolar division may take various angles towards the axis of the bulb, but these all depend on the extent that this organ is drawn out or retracted. A detached palpus cannot be accurately ascribed to any of the eremic species, apart from that of *L. dahli* which has a particularly short embolus, winding only once around the bulb (Figs 198–200).

The sympatric, arid populations in Israel can be distinguished from each other by distinct differences in the female genitalia, opisthosomal spination, the site and structure of the web, the pattern of colour of the adults and in particular that of the immature stages, and to some extent also by the season of adult male appearance (Levy & Amitai, 1983). The coastal population of *L. geometricus* can be distinguished from all the others by the coloured pattern of the adult and its characteristic, tufted egg sacs. Striking differences in the opisthosomal spination of the different species become distinct only in the mature female. The smooth, almost glabrous appearance of *L. pallidus* is due to its being clothed with widely dispersed, minute spines, exposing the

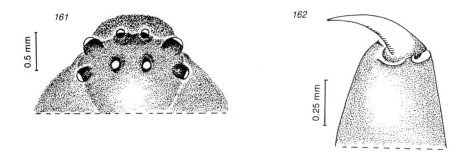

Figs 161–162: *Latrodectus tredecimguttatus* (Rossi, 1790); female
161. carapace, dorsal view of anterior part; 162. tip of left chelicera, inner view

integument, while the dull hue of *L. revivensis* is caused by slightly bent, thick spines scattered among appreciably smaller setae. The cone-like and bifurcate, unique spines covering the opisthosoma of *L. tredecimguttatus* are unlike the spines of any *Latrodectus* in this region (Figs 164 A–C).

The web of *L. tredecimguttatus* is found at the ground level often extending under stones, while those of *L. pallidus* and *L. revivensis* are spun among twigs of shrubs, well above the ground. The web site of the Israeli population of *L. hesperus* is under overhanging edges of stones and inside rock crevices, and *L. dahli* as known so far spins its web at the entrance of abandoned rodent burrows and fissures in the soil.

The female spider, on the slightest disturbance, moves to the deepest part of the web, beyond the egg sacs suspended ordinarily along the ceiling of the retreat. Apart from the peculiar egg sac of *L. geometricus*, all *Latrodectus* species of Israel have smooth, tough-coated, spherical to pear-shaped egg sacs. In all species the female constructs several egg sacs during its life-time. The colour of the egg sacs is usually white to creamy yellow and the eggs are similar in colour, sometimes yellow-orange. In summer hatching from eggs occurs within a week or two after oviposition. The newly hatched spiderlings are unpigmented at first. A few days later they undergo their first moult and gradually become coloured. On emergence from the egg sac the young of *L. tredecimguttatus* are already deep brown to black with rows of red spots, while the young of all other *Latrodectus* species in Israel are yellowish-white to light brown, each carrying the characteristic pattern of lines and blotches. The black ground-colour in both sexes of *L. revivensis* is gained only in the adult or partly in the penultimate stage. In the Israeli population of *L. hesperus* and in *L. dahli*, only the female becomes black while the adult male remains light.

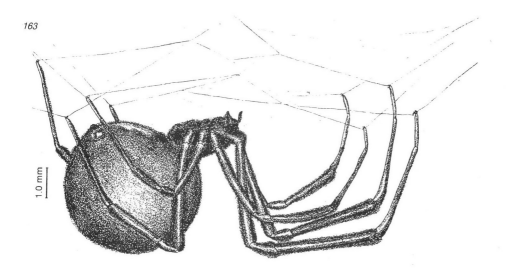

163

1.0 mm

Fig. 163: *Latrodectus tredecimguttatus* (Rossi, 1790); female,
lateral view of spider

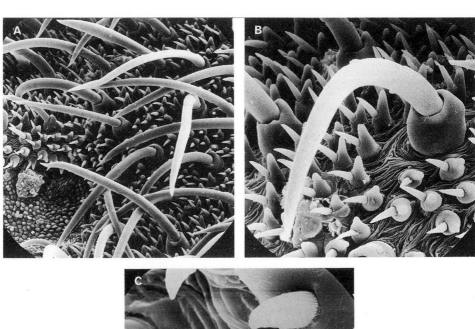

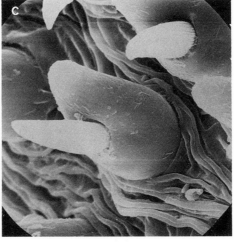

Fig. 164: *Latrodectus tredecimguttatus* (Rossi, 1790); female, spines on opisthosoma
A. Thick, long and arched spines rather closely dispersed among numerous cone-like, bifurcate, small spines (x 130); B. Enlargement of a single arched spine along with the coned, bifurcate, smaller ones (x 312); C. Enlargement of a small, cone-like, bifurcate spine (x 1300)

*Key to the Species of the **Latrodectus** in Israel*

Males:

1. Embolus of palpus coiling several times around cymbium (Fig. 174)   2
   – Embolus winding only once around bulb   **L. dahli** Levi
2. Embolus forms up to three coils around cymbium   3
   – Embolus circles four times or more around cymbium   **L. geometricus** C.L. Koch
3. Opisthosoma black with or without red or brownish spots on dorsum   4
   – Opisthosoma white or light brown   5
4. Opisthosoma dorsally with three rows of circular or crescent-shaped, usually separate spots, never forming a median continuous marking (Figs 165, 166) or opisthosoma evenly black with no spots; red spots on ventral surface, if present, are discontinuous (Figs 169–171)   **L. tredecimguttatus** (Rossi)
   – Opisthosoma dorsally with a median chain of large, irregular spots, and margins of black, ground colour deeply scalloped (Fig. 179); venter with light marking extending from epigastric furrow to spinnerets   **L. revivensis** Shulov
5. Ventral surface of opisthosoma with a large, white to yellowish shield-shaped marking
                                                            L. pallidus O.P.-Cambridge
   – Ventral surface with a large, deep red marking   **L. aff. hesperus** Chamberlin & Ivie

Females:

1. Opisthosoma white, yellowish or light brown   2
   – Opisthosoma black (when considering dark specimens living inside animal-breeding facilities — *cf. L. geometricus*)   3
2. White-creamy back of opisthosoma with rows of brown dots partly placed in small depressions formed by inner muscle attachments (Fig. 184); venter with large, light shield-shaped marking (Fig. 185); opisthosoma covered with few widely separated, small spines (Fig. 183 A)   **L. pallidus** O.P.-Cambridge
   – Light brown opisthosoma with two rows of black spots surrounded by dark lines and light blotches (Figs 203, 204); venter with large red marking; opisthosoma densely covered with fine long bristles   **L. geometricus** C.L. Koch
3. Ducts of spermathecae form only one lateral loop (requiring dissection to reveal them; Fig. 202)   **L. dahli** Levi
   – Ducts of spermathecae form two or more loops   4
4. Opisthosoma clothed with long bristles (Fig. 190); venter covered with a large red shield-shaped marking; ducts of spermathecae form two loops (Fig. 197)
                                                   L. aff. hesperus Chamberlin & Ivie
   – Opisthosoma clothed with strong spines; venter evenly black or with small, restricted markings; ducts of spermathecae form helices made of four loops (Figs 177, 182)   5
5. Surface of opisthosoma covered with long, slightly bent, evenly thick spines scattered among numerous, small cone-like and bifurcate spines (Figs 164 A–C); venter often traversed by a short red bar close behind epigastric furrow   **L. tredecimguttatus** (Rossi)
   – Surface of opisthosoma covered only with strong, thick, tapering spines dispersed among smaller ones of same kind (Figs 178 A–C); venter usually with a light, triangular marking
                                                            L. revivensis Shulov

Latrodectus tredecimguttatus (Rossi, 1790)

Figs 161–177

Aranea tredecimguttata Rossi, 1790, *Fauna Etrusca*, 2:136, pl. 9, fig. 10.

Latrodectus tredecimguttatus —. Walckenaer, 1806, *Histoire Naturelle des Aranéides*, pl. 1, fig. 5; Levy & Amitai, 1983, *Zool. J. Linn. Soc.* 77:46.

Length of male 3.2–5.5 mm, female 9.0–17.0 mm. Spines covering opisthosoma of female are of two kinds: long, arched, evenly thick spines, and numerous small, cone-like and bifurcate ones (Figs 164 A–C). Coloration of opisthosoma of adult female uniformly black on back and venter (Fig. 163), or venter partly traversed by a narrow red bar close behind the epigastric furrow. Occasional specimes are with

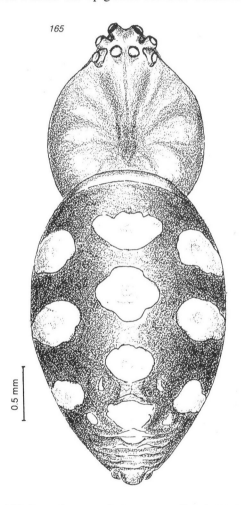

165

0.5 mm

Fig. 165: *Latrodectus tredecimguttatus* (Rossi, 1790); male, dorsal view of spider with ordinary form of markings on opisthosoma

166

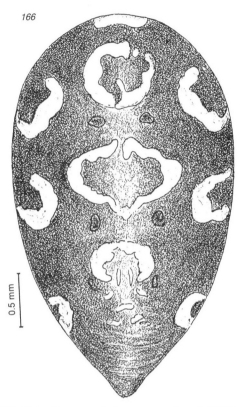

0.5 mm

Fig. 166: *Latrodectus tredecimguttatus* (Rossi, 1790); male,
opisthosoma, dorsal view, common variation

remnants of the colourful spotted pattern of the immature stage. The more common
dorsal opisthosomal pattern of the male and young stadia consists of a red frontal bar
and up to thirteen red spots arranged in three longitudinal rows (Fig. 165). The red
spots may be surrounded by distinct circular or semicircular borders (Fig. 166). The
posterior median spots may occasionally be partly joined, some of the posteriors may
be missing; rarely, some of the anteriors are joined across the upper surface (Fig. 167).
Sides of the opisthosoma when not black may be partly covered by drop-like extensions
of dorsal spots (Fig. 168). Venter, in ordinary form, displays an anterior red band,
sometimes discontinued at the middle, and a short posterior bar in front of the
spinnerets (Fig. 169); variations encountered are of bands slightly widened but never
forming a continuous median marking (Figs 170, 171).

Male Palpus: Cymbium of elaborate shape encircled by embolar coils (Figs 172–174).
Female Epigynum: Figs 175, 176. Spermathecae lie with their axes making an angle of
about 45° to each other, and ducts start looping from above the spermathecal body
(Fig. 177); ducts form four lateral coils, the smallest loop at helix tip, before it recurves
inwards on itself, is relatively slender (Fig. 177).

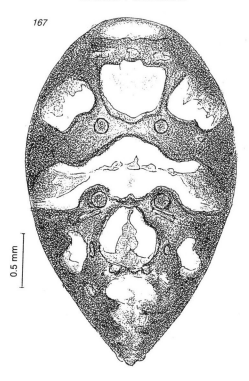

167

0.5 mm

Fig. 167: *Latrodectus tredecimguttatus* (Rossi, 1790); male,
opisthosoma, dorsal view, less common variation

Distribution: Mediterranean countries, possibly in Central Asia.

Israel: Throughout the country, scarce in the Mediterranean zone (1 to 11), abundant
in the Golan Heights (18), Jordan Valley (7) and the southern eremic parts (12 to 17),
and in Sinai (21, 22), taking into consideration great fluctuations in the density of the
populations.

Adult males are found from May to September, females mainly from May to November. Oviposition from June to October. A female may construct seven to eight egg
sacs at an interval of one to three weeks between sacs. Size of sacs about 14 to 17 mm
and some may contain over 450 eggs; size of an egg about one mm. In summer,
duration of time from oviposition to emergence may take about 40 days. In egg sacs
that have been produced towards onset of winter, the young apparently hibernate and
emerge only three-four months after oviposition. Outside the egg sac the males usually
undergo four or five moults to maturity, and the females six or seven moults. Young
females that emerged in early spring may reach maturity in about four months, while
those emerging towards the end of the summer may take 12 months or even longer to

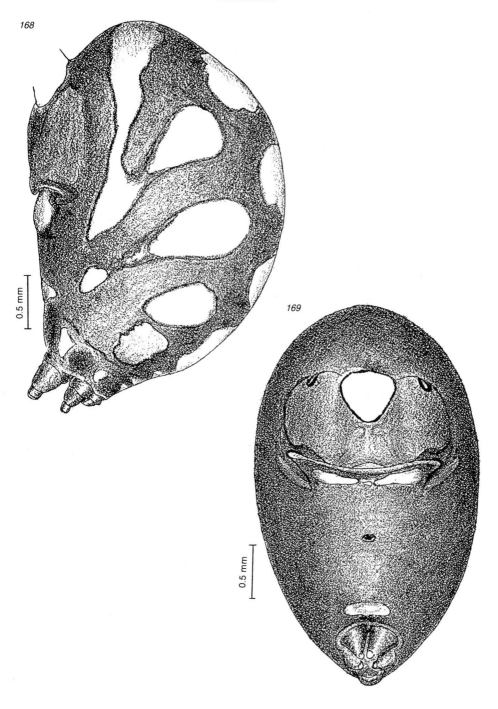

Figs 168–169: *Latrodectus tredecimguttatus* (Rossi, 1790); male, opisthosoma
168. lateral view, uncommon markings; 169. ventral view, ordinary form of markings

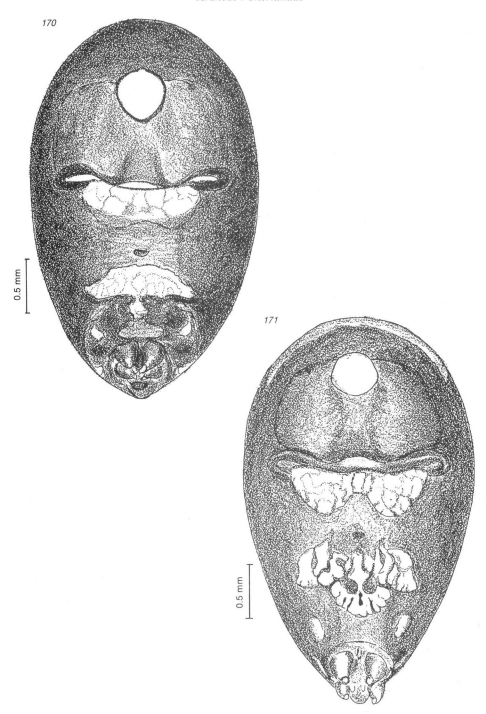

Figs 170–171: *Latrodectus tredecimguttatus* (Rossi, 1790); male, opisthosoma
170. ventral view, common variation; 171. ventral view, less common variation

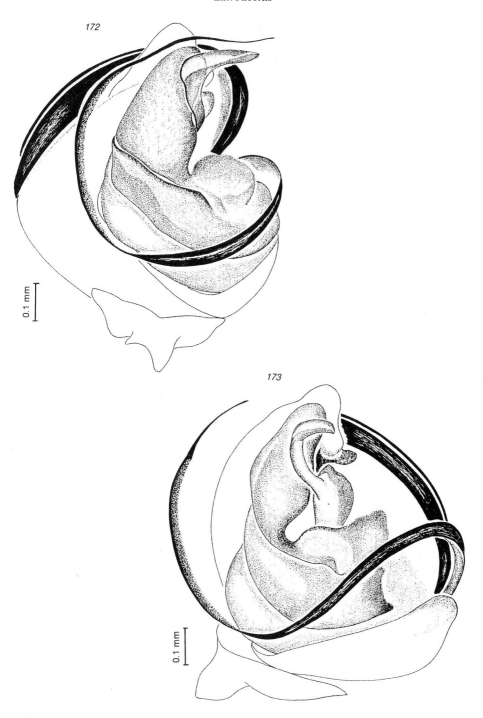

172

173

0.1 mm

0.1 mm

Figs 172–173: *Latrodectus tredecimguttatus* (Rossi, 1790); male, left palpus
172. mesal view; 173. ventral view

95

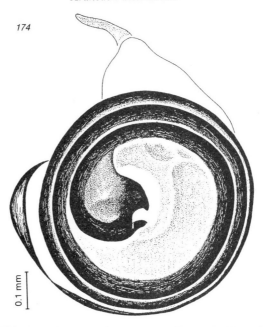

Fig. 174: *Latrodectus tredecimguttatus* (Rossi, 1790); male,
left palpus, apical view

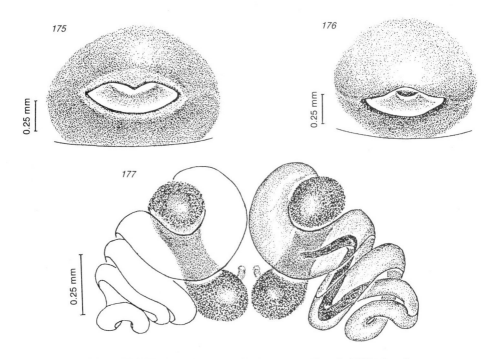

Figs 175–177: *Latrodectus tredecimguttatus* (Rossi, 1790); female
175. epigynum; 176. epigynum, variation; 177. inner spermathecae, dorsal view

reach maturity. The life-span of adult males in nature is usually only a few weeks while the females live for several months. Under laboratory conditions, females have been kept alive for over two years.

Latrodectus revivensis Shulov, 1948
Figs 178–182

Latrodectus revivensis Shulov, 1948, *Ecology*, 29:209, figs 1–12; Levy & Amitai, 1983, *Zool. J. Linn. Soc.* 77:51.

Length of male 5.7–7.5 mm, female 12.0–18.5 mm. Opisthosoma of female covered with medium-sized, thick and slightly bent spines scattered among appreciably smaller ones (Figs 178 A–C). Coloration of dorsal surface of adult females in spring to early summer populations, deep brown to black, and dull grey in late summer populations; venter often bears a small yellow triangular, sometimes elongated, anterior mark. Young stages are of pale hue. Opisthosomal coloration of male consists of a deeply scalloped black pattern on back with a median chain of, sometimes discontinued, large irregular spots (Fig. 179); sides are light and bear a row of elongated spots (Fig. 180). Venter displays a large, light triangular blotch extending from epigastric furrow to the spinnerets.

Male Palpus: As in *L. tredecimguttatus*.

Female Epigynum: Fig. 181. Position of spermathecae and starting point of loops of ducts as in *L. tredecimguttatus*. The ducts curving in four lateral coils are relatively thick, especially the smallest at the helix tip (Fig. 182).

Distribution: Southern Israel: Negev and ʿArava Valley (14–17).

Adult males are found in March–April and again in November, and adult females from February to about September. Females with egg sacs were collected from April to September. Usually there are three to four egg sacs in a web but eight and nine sacs were also encountered. Sizes of egg sacs are 10 to 18 mm and a sac may contain over one thousand eggs. Young emerge mainly in June–July. Males may reach maturity in little more than a month and females in about three months. Adult male may share the web of a penultimate female. The web is affixed to stems and branches of low shrubs usually about 30 to 40 cm above the ground. The web consists of an opaque, upper conical retreat extending below into a transparent funnel which is connected by a bridge-like network to a small irregularly meshed catching platform, fastened to the ground by a number of vertical threads. The vertical threads are covered at about two to five cm of their lower ends with viscid droplets. Carcasses, mainly of beetles are interwoven into the walls of the funnel.

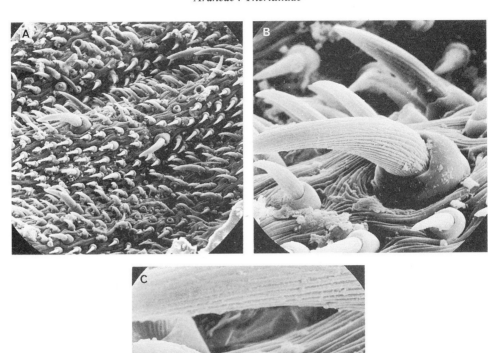

Fig. 178: *Latrodectus revivensis* Shulov, 1948; female, spines on opisthosoma
A. Medium-sized, thick and bent spines scattered among appreciably smaller ones
(x 130); B. Enlargement of a single arched spine along with smaller ones of same kind
(x 559); C. Enlargement of a single, small spine (x 1560)

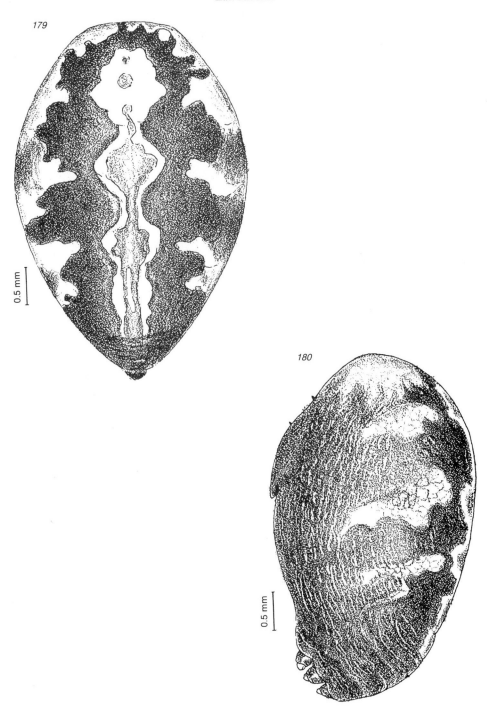

Fig. 179–180: *Latrodectus revivensis* Shulov, 1948; male, opisthosoma
179. dorsal surface; 180. lateral view

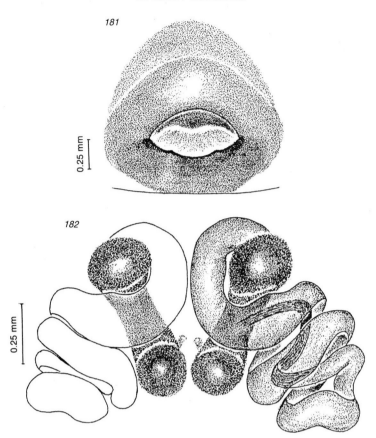

Figs 181–182: *Latrodectus revivensis* Shulov, 1948; female
181. epigynum; 182. inner spermathecae, dorsal view

Latrodectus pallidus O.P.-Cambridge, 1872
Figs 183–189

Latrodectus pallidus O.P.-Cambridge, 1872, *Proc. zool. Soc. Lond.* 1872:287; Levy & Amitai, 1983, *Zool. J. Linn. Soc.* 77:47.

Length of male 3.5–5.5 mm, female 11–13 mm. Integument of female opisthosoma covered only by widely dispersed, small spines (Figs 183 A–B). The opisthosoma of the young, as well as the adult female dorsally, is creamy white with brown dots arranged in rows, and the sides are sometimes suffused with yellowish to light brown stripes (Fig. 184). A large white to yellow shield-shaped mark covers the entire ventral space between the epigastric furrow and the spinnerets, and four to six light, small conspicuous spots surround the spinnerets (Fig. 185). The male bears dorsally on the opisthosoma a pattern of dark lines surrounding black spots and extending down the sides,

100

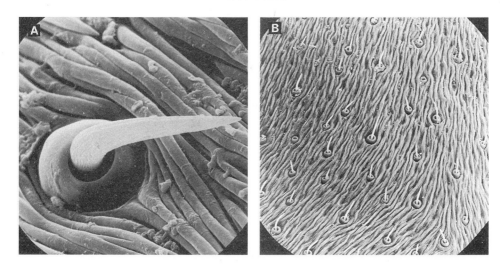

Fig. 183: *Latrodectus pallidus* O.P.-Cambridge, 1872; female, spines on opisthosoma
A. Enlargement of a single spine (x 1300); B. Integument covered only by widely dispersed, small spines (x 130)

but often only the black spots are visible on the pearly-white background (Fig. 186); the ventral surface is covered by a large shield-shaped yellow marking.

Male Palpus: As in *L. tredecimguttatus*.

Female Epigynum: Fig. 187. Angle of spermathecal bodies towards each other and apical starting point of loops resemble that of *L. tredecimguttatus*, but the ducts form only three major lateral loops (Fig. 188).

Distribution: Asian Russia, Iran, Turkey (Anatolia), Israel, Libya, presumably in Syria, Jordan and Egypt (Sinai) but there are no explicit records.

Israel: Formerly in the Valley of Yizre'el (5) and central Coastal Plain (8), but at present along the eastern parts in the Jordan Valley (7), Judean Desert and Dead Sea area (12, 13) throughout the Negev to Elat (14–17).

Adult females are found nearly the year round. Since certain pigments wash out, immediately on preservation, the adult males of *L. pallidus* can be separated with certainty from males of *L. aff. hesperus* only while alive. Authenticated records of males, also of those taken in the field together with females, are from February, June, August and October. Oviposition is in the warmer months, primarily in late summer. Size of egg sac is about 12 mm and each may contain 40 to over 200 eggs. The web is spun on shrubs at heights of up to 60 cm from the ground, resembling in general the structure of *L. revivensis*. The coned retreat tapers obliquely above the catching platform (Fig. 189). Remnants of encased ants are particularly abundant, but scorpions, isopods and other arthropods, as well as dry leaves and small stones, cover the upper parts of the cone-shaped retreat. Awaiting prey on the catching platform and spinning activities take place at night or in the early morning hours.

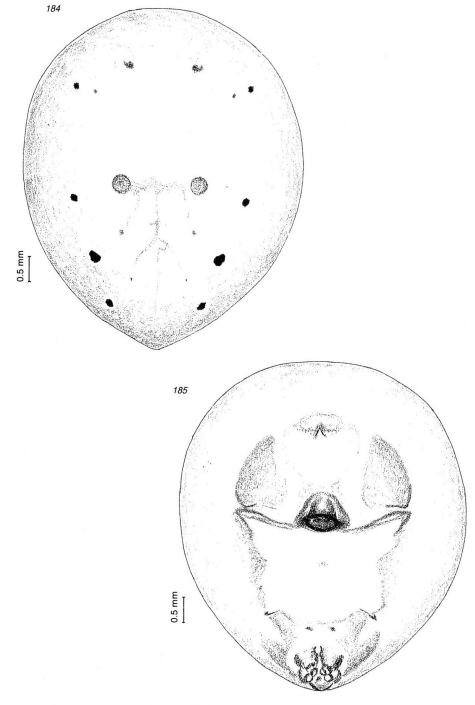

Fig. 184–185: *Latrodectus pallidus* O.P.-Cambridge, 1872; female, opisthosoma
184. dorsal surface; 185. ventral surface

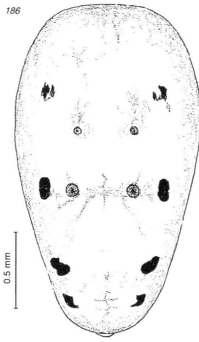

Fig. 186: *Latrodectus pallidus* O.P.-Cambridge, 1872; male,
opisthosoma, dorsal surface

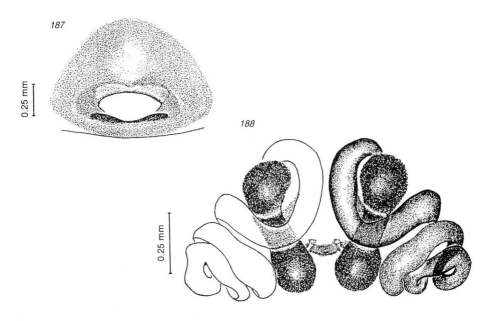

Figs 187–188: *Latrodectus pallidus* O.P.-Cambridge, 1872; female
187. epigynum; 188. inner spermathecae, dorsal view

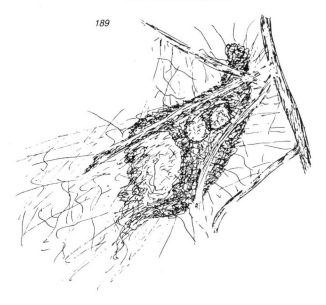

Fig. 189: *Latrodectus pallidus* O.P.-Cambridge, 1872; coned retreat in web,
two egg sacs discernible (diagrammatic)

Latrodectus ?hesperus Chamberlin & Ivie, 1935

Figs 190–197

Latrodectus mactans hesperus Chamberlin & Ivie, 1935, *Bull. Univ. Utah, Biol. Ser.* 3:15, figs 1,
　4, 6–14, 21, 23–33.
Latrodectus hesperus —. Kaston, 1970, *Trans. San Diego Soc. nat. Hist.* 16:39; Levy & Amitai,
　1983, *Zool. J. Linn. Soc.* 77:53 (as *L.* aff. *hesperus*).

Length of male 3.0–3.7 mm, female 11.7–19.1 mm. Opisthosoma of female clothed
only with long, attenuated bristles (Fig. 190). Adult female is shiny black with
deep-red, large shield-shaped marking on venter; the red pigment washes out immediately on preservation. Young female is dorsally brown with a median row of large
spots and marginal rows of black, oval spots surrounded by dark lines, and light
blotches extending down the sides (Fig. 191); the venter is covered by a large red
marking (Fig. 192). Opisthosomal coloration of adult male is yellow to light brown
with two rows of black spots on dorsum; dark line surrounds the black spots and
extends down the sides (Figs 193, 194); venter bears a large red marking (Fig. 195).
Male Palpus: As in *L. tredecimguttatus.*
Female Epigynum: Fig. 196. Reddish black dumb-bell spermathecae lie parallely and
their ducts start winding medially around the shaft of the spermathecae, at a point
below the tip (Fig. 197); the ducts form two lateral loops outside the spermathecae
(Fig. 197).

Distribution: Israel (for Middle East population only).

Israel: Jordan Valley (7), Judean Desert (12) and Dead Sea area (13) to the Northern and Central Negev (15, 17).

Adult males were found only in the winter months, November to January, always together with immature females. Mature females occur mainly from March to June, some were also found in October and December. Oviposition is in early spring. Size of egg sacs is about 15 mm and each may contain several hundred eggs. The web is spun under overhanging edges of cliffs and like that of *L. tredecimguttatus* comprises an irregular catching network and a hardly outlined inner retreat concealed in a recess in the rock.

There may be subtle differences between the Middle East and the American populations of *L. hesperus* but whether these provide criteria for their discrimination or whether the two belong to the same species cannot be judged at present. The Middle East population is referred to therefore as *L.* aff. *hesperus*.

190

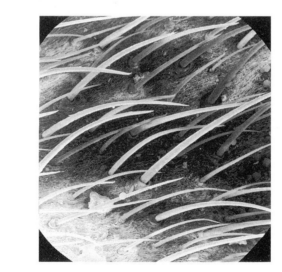

Fig. 190: *Latrodectus ?hesperus* Chamberlin & Ivie, 1935; female,
opisthosoma bearing only long, attenuated bristles (x 130)

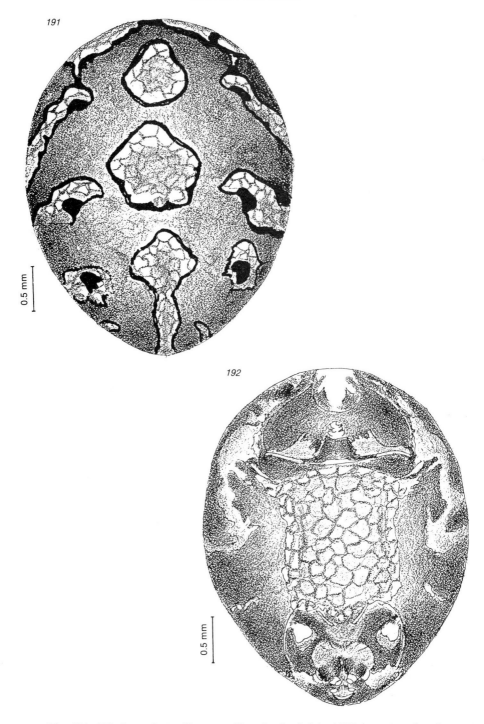

Figs 191–192: *Latrodectus ?hesperus* Chamberlin & Ivie, 1935; immature female
191. opisthosoma, dorsal surface; 192. opisthosoma, ventral surface

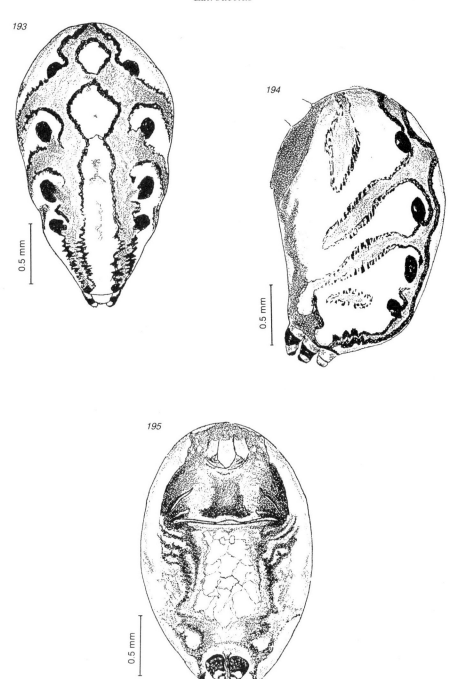

Figs 193–195: *Latrodectus ?hesperus* Chamberlin & Ivie, 1935; male, opisthosoma
193. dorsal view; 194. lateral view; 195. ventral view

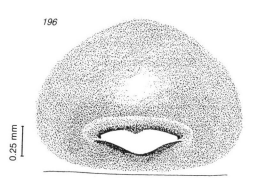

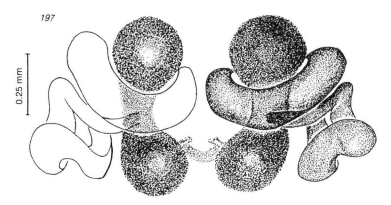

Figs 196–197: *Latrodectus ?hesperus* Chamberlin & Ivie, 1935; female
196. epigynum; 197. inner spermathecae, dorsal view

Latrodectus dahli Levi, 1959
Figs 198–202

Latrodectus dahli Levi, 1959, *Trans. Am. microscop. Soc.* 78:42, figs 11, 12; Levy & Amitai,
1983, *Zool. J. Linn. Soc.* 77:55.

Length of male 2.9 mm, female 15 mm. Opisthosoma of female somewhat sparsely
covered by long bristles dispersed among shorter ones similarly shaped. Coloration of
adult female is black with only a fine light line on venter, close behind the epigastric
furrow and a small spot in front of the spinnerets. Young stages as well as the adult
male are light yellow to white with three to five pairs of black spots on dorsum
surrounded by brown blotches extending down the sides and anteriorly; along venter
the male displays a large marking.

Male Palpus: Embolus forms only one coil around cymbium (Figs 198–200).

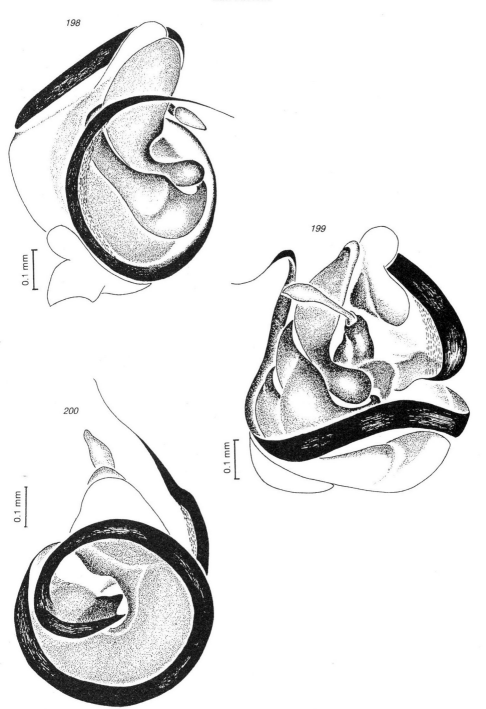

Figs 198–200: *Latrodectus dahli* Levi, 1959; male, left palpus
198. mesal view; 199. ventral view; 200. apical view

Female Epigynum: Fig. 201. Angle of spermathecal bodies towards each other and medial starting point of loops resemble that of *L.* aff. *hesperus*, but the ducts form only one lateral loop (Fig. 202).

Distribution: Southern Uzbekistan, Iran (Bushire), Island of Socotra, southern Israel. Israel: Ze'elim (15) and Makhtesh Gadol (17).

An adult male was found in February and a female in April.

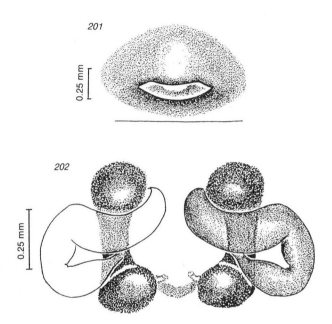

Figs 201–202: *Latrodectus dahli* Levi, 1959; female
201. epigynum; 202. inner spermathecae, dorsal view

Latrodectus geometricus C.L. Koch, 1841
Figs 203–207

Latrodectus geometricus C.L. Koch, 1841, *Die Arachniden* 8:117, pl. 284, fig. 684; Roewer, 1942, *Katalog der Araneae*, 1:425; Bonnet, 1957, *Bibliographia Araneorum* 2(3):2368; Levi, 1959, *Trans. Am. microscop. Soc.* 78:21; Levy & Amitai, 1983, *Zool. J. Linn. Soc.* 77:56.

Length of male 3.4 mm, female 8.1–8.9 mm. Opisthosoma of mature female clothed with fine, very long hair-like bristles, longer than in all other *Latrodectus* of Israel. Light brown to black variants are encountered. In the light ones a pattern of black spots and light discontinued median blotches cover the back; blotches partly taper down the sides (Figs 203, 204); venter displays a large, deep red, shield-shaped marking. Pattern of male resembles that of *L.* aff. *hesperus* (Figs 193–195).

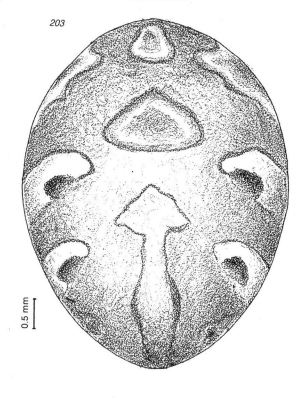

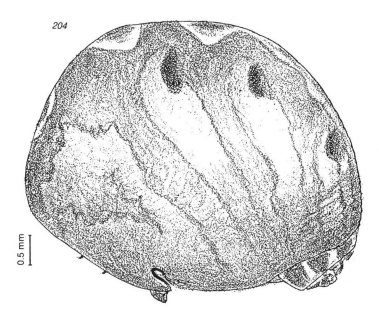

Fig. 203–204: *Latrodectus geometricus* C.L. Koch, 1841; female, opisthosoma
203. dorsal surface; 204. lateral view

Male Palpus: As in *L. tredecimguttatus* with embolus winding four times or more around cymbium.

Female Epigynum: Fig. 205. Small parallel spermathecal bodies, give rise medially to lateral ducts forming four tight coils (Fig. 206).

Distribution: Cosmotropical; carried by man around the world.

Israel: Along the Coastal Plain (8, 9).

Adults were collected around man-made habitats in March and December. Females with two to five egg sacs each were taken in those months. Inside animal-breeding rooms, adults are possibly found throughout the year. Their tufted egg sacs (Fig. 207) are about ten mm in size and contain about 100 eggs.

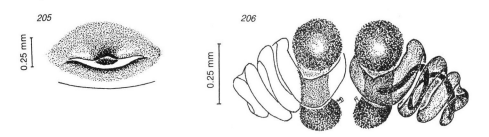

Figs 205–206: *Latrodectus geometricus* C.L. Koch, 1841; female
205. epigynum; 206. inner spermathecae, dorsal view

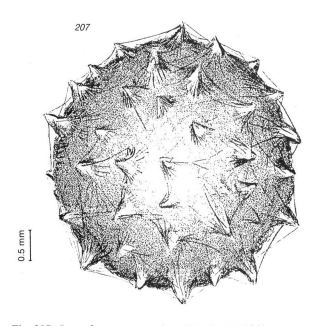

Fig. 207: *Latrodectus geometricus* C.L. Koch, 1841; egg sac

Genus ARGYRODES Simon, 1864

Histoire naturelle des Araignées (Aranéides), Paris:253.

Figs 208–213

Type-species: *Linyphia argyrodes* Walckenaer, 1841.

Theridiids with pronounced sexual dimorphism. Carapace of male with raised eye region or projecting clypeal boss (Fig. 208), or with clypeus indented by deep cleft below median eyes (Fig. 224). Carapace of female moderate to flat (Figs 211, 227). Arrangement of eyes in males varies with modifications of carapace, but ususally anterior-lateral eyes and posterior-laterals touching (Figs 209, 210, 225, 226). Chelicerae armed with promarginal and retromarginal teeth. Legs often long (Fig. 213): first pair longest, third shortest. Opisthosoma sometimes higher than long (Figs 230, 231) but often extends beyond spinnerets and ends with small prominences (Figs 213–216), or vermiform bearing fleshy processes. Colulus distinct, usually with two seate.

Argyrodes spiders are usually found in association with webs of other spiders, mostly Araneidae, but also with Pholcidae, Linyphiidae, Agelenidae and other Theridiidae. They may live in the tangled webs of these without building a web of their own. Many are referred to as kleptoparasites, feeding on their host's prey, but some catch young of their host and are thus considered predators. The kleptoparasites scavange very small prey that become caught in the host's web or move on to larger prey already killed and enswathed in silk by the host. They may sever the silk threads attaching stored prey to the host web and drag it away with their own threads, or join the host feeding on the same prey. *Argyrodes* spiders usually hang in the web upside-down. They may move slowly, using their long front pair of legs as feelers, and tend to drop instantly to the ground, leaving a line attached, when disturbed; the host only infrequently reacts to the presence of *Argyrodes* in its web. *Argyrodes* is not host-specific and may move from one host to another.

Males of *Argyrodes* have a special clypeal gland below their ocular protuberance. It plays a rôle during copulation in which the female is holding firmly on to the projection on the male's forehead with her chelicerae. The egg sacs are of various forms, often pear- or urn-shaped cases with thin, tough walls suspended by a fine stalk from the host web or inside a web of their own.

Argyrodes spiders occur in the warmer parts of the world, being most abundant in the tropics. Numerous species have been described but apparently none is known to be cosmotropical. Only *A. argyrodes* occurs in southern Europe and it was also collected once, over one hundred years ago, in Israel (O.P.-Cambridge, 1872). *Argyrodes syriaca* is now commonly found in north and central Israel, while a third species, *A. longicaudata* (O.P.-Cambridge, 1872), with a peculiar, exceedingly tapered opisthosoma (Levy, 1985a) is still known only by the male holotype from Beirut, Lebanon.

Key to the Species of **Argyrodes** in Israel

1. Male with a clypeal projection separated by a deep cleft from swollen region bearing the median eyes (Figs 224–226). Female with opisthosoma higher than long, ending with a single tip (Fig. 231)   **A. argyrodes** (Walckenaer)

- Male with a projecting clypeal boss and a parallel cephalic hump that does not bear the eyes (Figs 208–210). Opisthosoma of female extended, longer than high, often ending with four tips (Figs 215, 216, 218)   **A. syriaca** O.P.-Cambridge

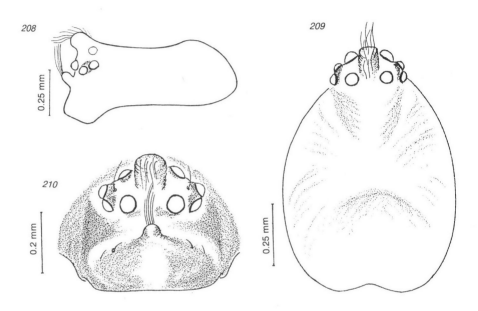

Figs 208–210: *Argyrodes syriaca* O.P.-Cambridge, 1872; male
208. prosoma, lateral view; 209. carapace, dorsal view; 210. carapace, frontal view

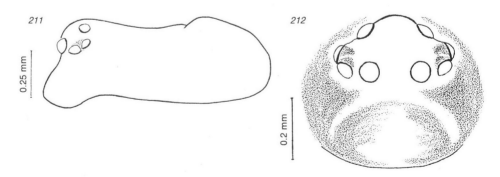

Figs 211–212: *Argyrodes syriaca* O.P.-Cambridge, 1872; female
211. prosoma, lateral view; 212. carapace, frontal view

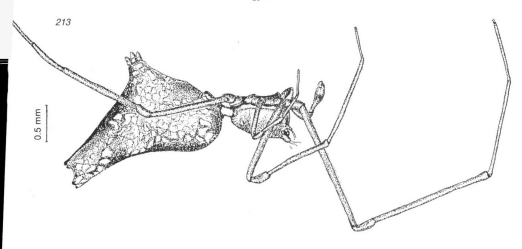

Fig. 213: *Argyrodes syriaca* O.P.-Cambridge, 1872; male,
lateral view of spider

Argyrodes syriaca O.P.-Cambridge, 1872
Figs 208–223

Argyrodes syriaca O.P.-Cambridge, 1872, *Proc. zool. Soc. Lond.* 1872:279, pl. 13, fig. 10; Levy,
    1985a, *J. Zool. Lond.* 207:103.

Length of male 2.4–4.6 mm, female 3.7–5.3 mm. Coloration of carapace red-brown to
black (Figs 208–212). Male without visible stridulatory ridges. Legs yellowish, partly
brown-banded. Opisthosoma varies greatly in size, triangular in profile, obtusely
produced behind and with four, short, divergent projections (Figs 213–216); brown to
black mottled with silver, or silvery almost throughout with only a black, mid-dorsal
line and a few black stripes (Figs 213, 217, 218).
Male Palpus: Relatively small. Median apophysis slightly concave, apical portion
rough (Figs 219, 220); embolar division with two processes, one rather large and
pointed (Figs 219, 220); outline of soft, fleshy conductor hardly visible.
Female Epigynum: Usually covered by hard, resinous-like material. Exposed, sclerotized plate beneath with only a tongue-like, narrow, median bulge accompanied by two
light tubes (Fig. 221); internal organs sometimes discernible through integument.
Thick tubes inside partly concealed by the large, compact spermathecal bodies (Fig.
222).
Distribution: Israel, Lebanon.
Israel: Upper Galilee (1), northern and central Coastal Plain (4, 8) to Samaria (6) and
the Judean Hills (11).

115

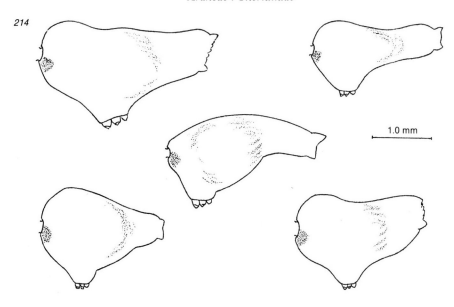

Fig. 214: *Argyrodes syriaca* O.P.-Cambridge, 1872; males, opisthosoma, lateral view; variations

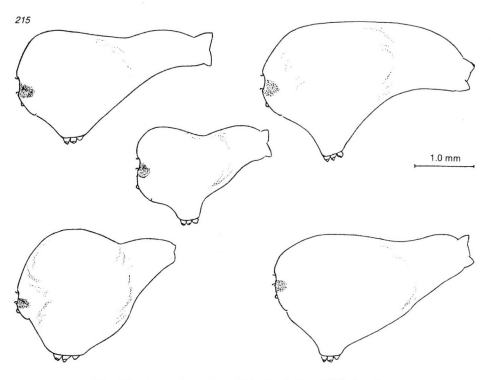

Fig. 215: *Argyrodes syriaca* O.P.-Cambridge, 1872; females, opisthosoma, lateral view; variations

216

Fig. 216: *Argyrodes syriaca* O.P.-Cambridge, 1872; female,
posterior tip of opisthosoma from behind

217

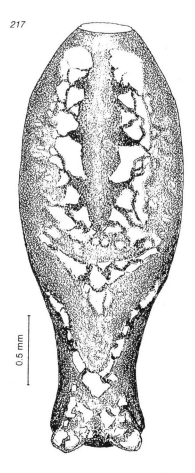

Fig. 217: *Argyrodes syriaca* O.P.-Cambridge, 1872; male,
opisthosoma, dorsal surface

117

Adult males are found in April–May and adult females from April to July. Most specimens were collected from underneath water-dripping *Adiantum* ferns. Some were taken from webs of Linyphiidae, Pholcidae and Uloboridae. In Lebanon *A. syriaca* was first described on webs of *Cyrtophora* (Araneidae). Females with egg cases suspended each by 15–20 mm long twined cords were found under dense humid vegetation in July. The egg cases, about six mm in length have a special orifice at the bottom for the young to emerge. Two kinds of egg cases: a light, fuzzy nearly transparent type and a yellowish papery and opaque kind (Fig. 223), were found hanging side by side accompanied by adult females of one and the same population.

218

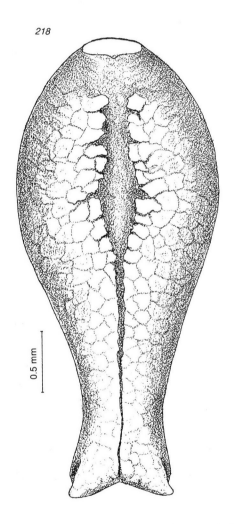

0.5 mm

Fig. 218: *Argyrodes syriaca* O.P.-Cambridge, 1872; female, opisthosoma, dorsal view

118

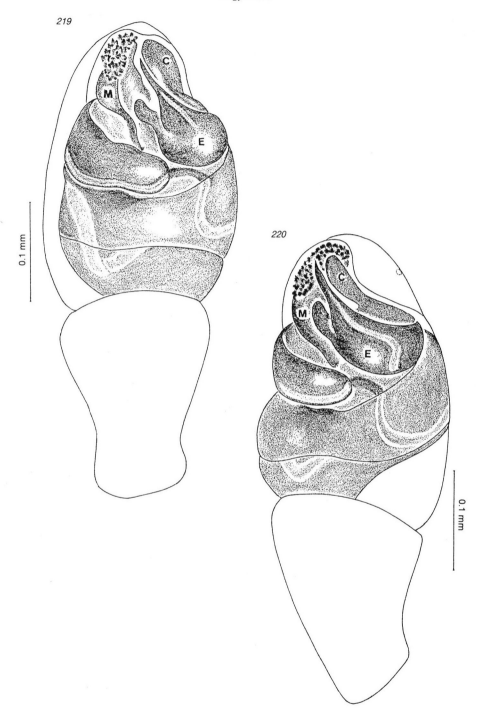

Figs 219–220: *Argyrodes syriaca* O.P.-Cambridge, 1872; male, left palpus
219. ventral view; C – conductor, E – embolus, M – median apophysis; 220. retrolateral view

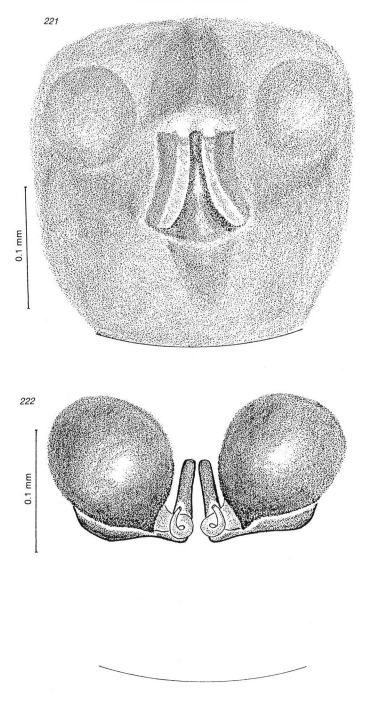

Figs 221–222: *Argyrodes syriaca* O.P.-Cambridge, 1872; female
221. epigynum; 222. inner spermathecae, dorsal view

Fig. 223: *Argyrodes syriaca* O.P.-Cambridge, 1872; egg cases

Argyrodes argyrodes (Walckenaer, 1841)
Figs 224–237

Linyphia argyrodes Walckenaer, 1841, *Histoire naturelle des Insectes Aptères*, Paris, 2:282.
Argyrodes argyrodes —. Simon, 1881, *Les Arachnides de France* 5:16; Roewer, 1942, *Katalog der Araneae*, 1:430; Exline & Levi, 1962, *Bull. Mus. comp. Zool. Harv.* 127:134; Levy, 1985a, *J. Zool. Lond.* 207:99.

Length of male 3.3–4.0 mm, female 3.0–4.1 mm. Carapace dark brown (Figs 224–229). Male with almost indistinct stridulatory ridges, posteriorly. Legs yellowish. Opisthosoma of male triangular with a dark, dorsal median band and large, partly contiguous, silvery patches on back and sides (Fig. 230); venter black, traversed by a few small silvery markings. Female with a high, cone-shaped opisthosoma, bright silver, apart from a short, black mid-dorsal line and a black spot on posterior tip (Figs 231, 232); venter and pedicel in front black.

Male Palpus: Relatively large. Cymbium bilobed (Fig. 233). Median apophysis elongated, tip triangular (Figs 234, 235); embolar division with three curved processes (Fig. 234); fleshy conductor carried on a fine inner shaft and crowned by an almost indistinct, transparent membrane, bears a crescentic process placed directly above embolar process (Figs 234, 235).

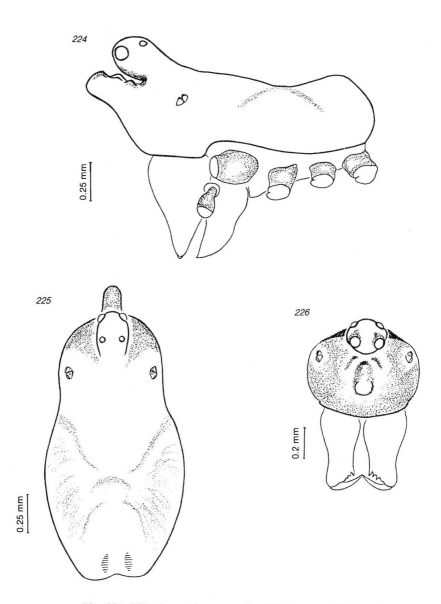

Figs 224–226: *Argyrodes argyrodes* (Walckenaer, 1841); male
224. prosoma, lateral view (legs omitted); 225. carapace, dorsal view;
226. carapace and chelicerae, frontal view

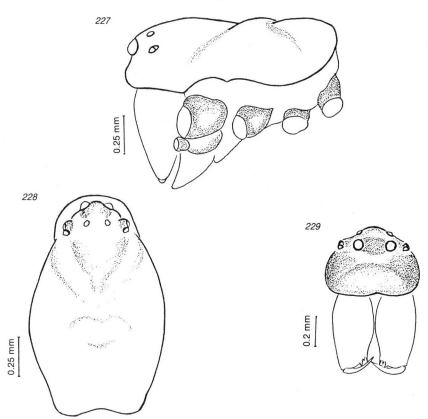

Figs 227–229: *Argyrodes argyrodes* (Walckenaer, 1841); female
227. prosoma, lateral view (legs omitted); 228. carapace, dorsal view;
229. carapace and chelicerae, frontal view

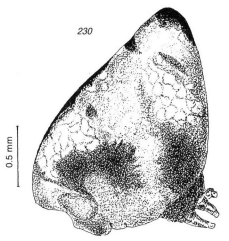

Fig. 230: *Argyrodes argyrodes* (Walckenaer, 1841); male,
opisthosoma, lateral view

123

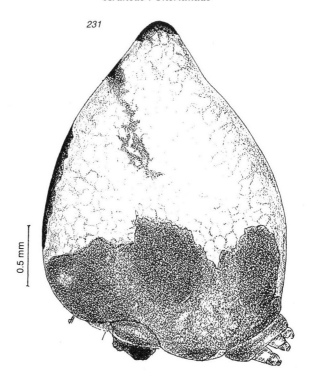

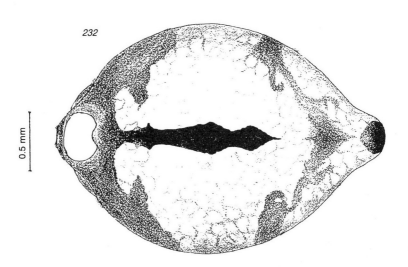

Figs 231–232: *Argyrodes argyrodes* (Walckenaer, 1841); female, opisthosoma
231. lateral view; 232. frontal surface

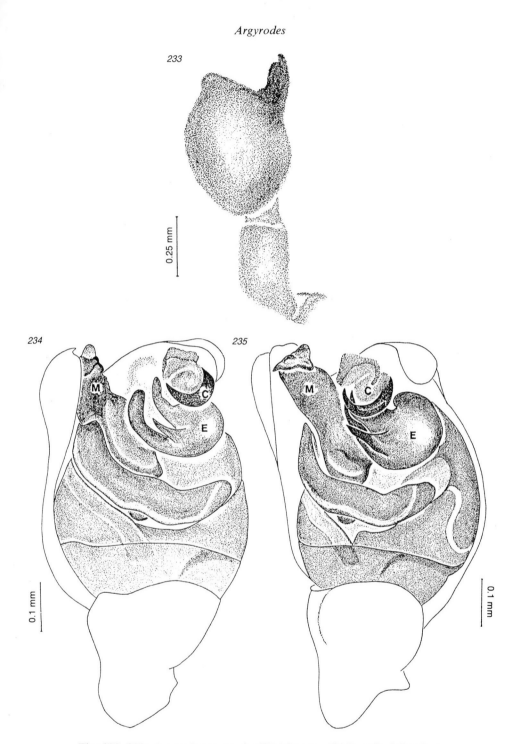

Figs 233–235: *Argyrodes argyrodes* (Walckenaer, 1841); male, left palpus
233. dorsal view of distal segments; 234. mesal view; C — conductor,
E — embolus, M — median apophysis; 235. ventral view

125

Female Epigynum: Often covered by a thick clot of resinous-like material. Sclerotized plate underneath displays two partly coiled, large orifices (Fig. 236); rims of orifices partly indistinct. Internal spermathecae form two brown, compact bodies, each giving rise to two, partly parallel, thick, short tubes (Fig. 237).

Distribution: Mediterranean countries, Canary Islands, Seychelles Islands, tropical atoll of Aldabra in the western Indian Ocean.

Israel: Tiberias, near Lake Kinneret (O.P.-Cambridge, 1872).

O.P.-Cambridge (1872:279) found two dozen specimens of *A. argyrodes* in April, 1865, in webs of *Cyrtophora* (Araneidae). Despite extensive searches not a single specimen has been found again in Israel. Since O.P.-Cambridge reported finding many *A. argyrodes* in Israel, while in Lebanon he seemed to encounter only *A. syriaca* which at present is common in Israel, perhaps the latter did replace *A. argyrodes* in Israel in the last century.

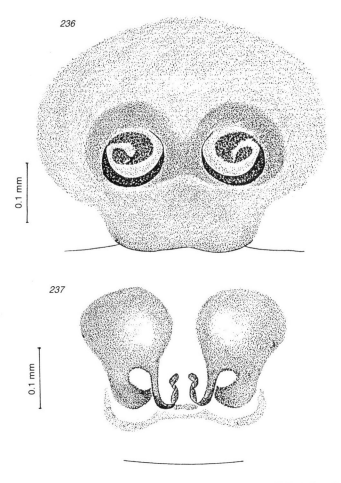

Figs 236–237: *Argyrodes argyrodes* (Walckenaer, 1841); female
236. epigynum; 237. inner spermathecae, dorsal view

126

Genus EPISINUS Latreille, 1809

Genus EPISINUS Latreille, 1809

Genera Crustaceorum et Insectorum, Paris 4:371

Figs 238–241

Type-species: *Episinus truncatus* Latreille, 1809.

Theridiids with total body length to about 6.0 mm (Figs 238, 248). Carapace longer than wide; thoracic region about as high as eye region (Fig. 239), often with a deep,

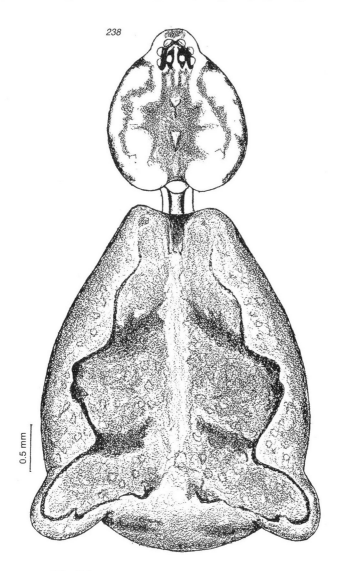

Fig. 238: *Episinus israeliensis* Levy, 1985; female,
dorsal view of spider

median longitudinal depression. Median eyes closer to laterals than to each other (Fig. 240). Lateral eyes often slightly separated. Chelicerae with one promarginal tooth (Fig. 241) or no teeth. Legs long, usually first pair longest, third shortest. Opisthosoma widest behind middle and usually modified with humps near posterior end (Fig. 238). Colulus with two setae.

Episinus spiders in Israel live in moist places with dense vegetation and are considered rare. They merely spin a few fine, hardly visible threads, and drop to the ground at the slightest disturbance. When not moving, the *Episinus* spiders hang upside down with slightly diverging, outstretched legs. The foreleg is joined by the second one, while the third leg stretches along the hind leg, so the spider looks as if it has only two legs on each side. Special sticky threads are attached to the ground for entrapping ants.

Episinus spiders in Israel are easily recognized by the shape of the modified opisthosoma and the relatively large palpi of the males. There is, however, great structural similarity with species occurring in Europe and North Africa and very careful study of the palpal organs and internal organs of the females is essential for separating the different species (Levy, 1985a). A few dozen species of *Episinus* are known from throughout the world, mainly from the tropics. Two species in Israel, one is known so far by the female only.

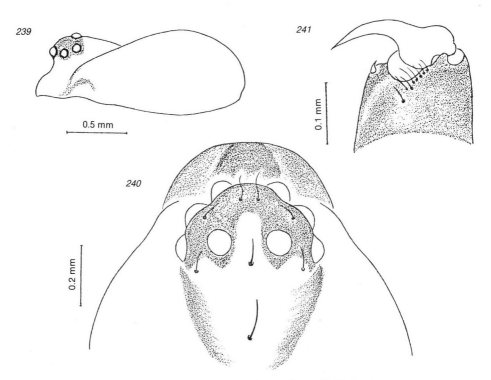

Figs 239–241: *Episinus israeliensis* Levy, 1985; female
239. prosoma, lateral view; 240. carapace, dorsal view of anterior part;
241. tip of left chelicera, inner view

128

*Key to the female species of **Episinus** in Israel*

1.   Opisthosoma dorsally flat (Fig. 242), posterior part with rounded, lateral protuberances (Fig. 238); epigynal plate relatively large (Fig. 246), spermathecae as in Fig. 247

   E. israeliensis Levy

–   Opisthosoma dorsally round (Fig. 249), posterior humps indistinct (Fig. 248); epigynal plate very small (Fig. 250), spermathecae as in Fig. 251   **E. fontinalis** Levy

Episinus israeliensis Levy, 1985
Figs 238–247

Episinus israeliensis Levy, 1985a, *J. Zool. Lond.*, 207:92.

Length of male 3.8 mm, female 4.3–5.8 mm. Coloration of carapace light, almost transparent, with a dorsal, broad, black median band, a few black markings on sides, and black margins (Fig. 238). Sternum black, somewhat shiny, and with a light, median longitudinal marking. Legs light, mottled with a few black spots; cymbii of male palpi blackish. Opisthosoma elongated, dorsally flat and in female with two conspicuous humps projecting on sides posteriorly (Figs 238, 242); humps less protuberant in male; dorsal surface marked by a blackish, folium-like pattern encircled by white and yellow margins (Fig. 238); sides of opisthosoma mottled with white dots and reddish black streaks; venter, in front with a black, shiny, broad band and space between epigastric furrow to spinnerets, marked by two, black, longitudinal, converging bands; spinnerets, light brown.

Male Palpus: Median apophysis slender with hooked tip (Figs 243–245); dark, cup-shaped tip of conductor slightly pointed; tegulum projects distally as a thin platelet, truncated apically (Fig. 245); close to tips of median apophysis and conductor projects a black, strong, pointed accessory apophysis (Figs 243, 245).

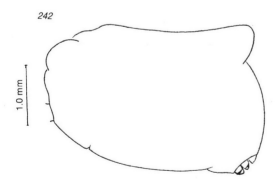

242

1.0 mm

Fig. 242: *Episinus israeliensis* Levy, 1985; female, opisthosoma, lateral view

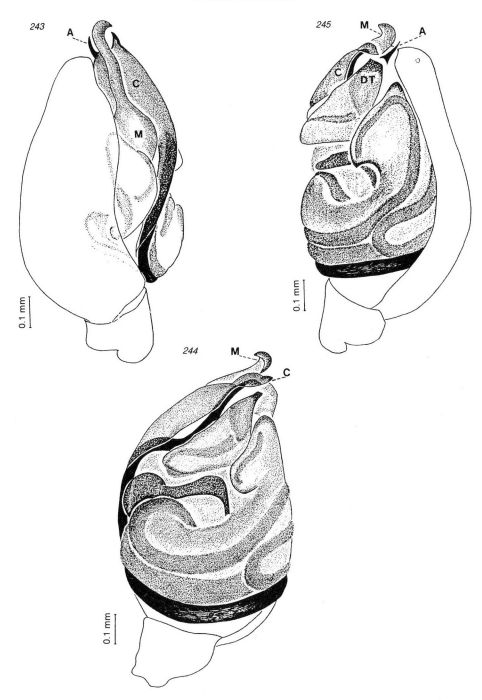

Figs 243–245: *Episinus israeliensis* Levy, 1985; male, left palpus
243. mesal view; A – accessory apophysis, C – conductor, DT – distal tegular projection,
M – median apophysis; 244. ventral view; 245. retrolateral view

Female Epigynum: Large, with an elongated, distinct central orifice bordered proximally by a dark rim (Fig. 246); internal bodies of spermathecae discernible above proximal rim of central orifice (Fig. 246). Dark, sclerotic spermathecal bodies and ducts almost entirely concealed by large, fleshy folds (Fig. 247).

Distribution: Israel.

Israel: Regba in the northern Coastal Plain (4), Samaria (6) Nahshon (10) and Jerusalem (11).

Adults are found from April to August. During the day the spider hides, hanging on a few threads inside dense vegetation and only late at night takes its hunting posture, suspended with outstretched legs. A thread about 10–15 cm long extends obliquely up and outside from each of the hind pairs of legs, and two very short threads, 2–3 cm in length are fastened to the ground and are held by the diverging front pairs of legs. The lowest portion of these threads, about 3–4 mm only is covered with sticky droplets.

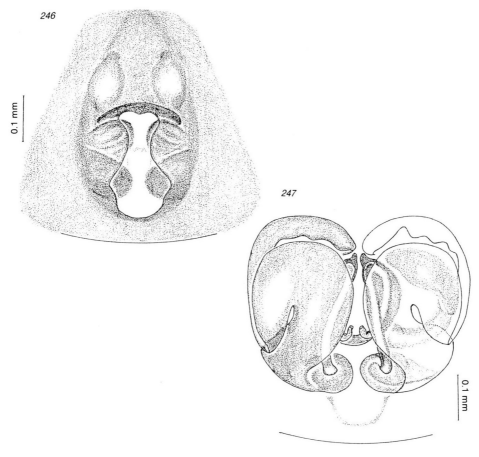

Figs 246–247: *Episinus israeliensis* Levy, 1985; female
246. epigynum; 247. inner spermathecae, dorsal view

Episinus fontinalis Levy, 1985
Figs 248–251

Episinus fontinalis Levy, 1985a, *J. Zool. Lond.*, 207:94.

Adult male unknown. Length of female 4.7 mm. Carapace black dorsally along median depression to clypeus, sides light with a dark zigzag line, margins black (Fig. 248). Sternum dark throughout. Legs light with black annulations. Opisthosoma rounded, widest behind middle, without posterior humps (Fig. 249); dorsal surface with a distinct, dark, folium-like pattern (Fig. 248); sides and venter mottled with black and white markings; spinnerets encircled with a dark, broad band.

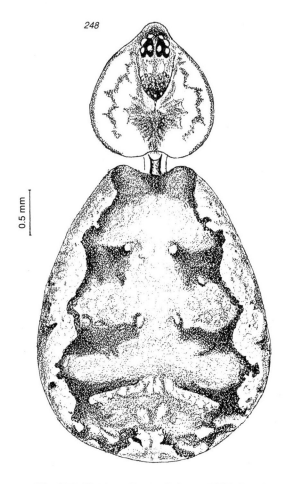

Fig. 248: *Episinus fontinalis* Levy, 1985; female,
dorsal view of spider

Female Epigynum: Small, with a rounded central orifice bordered on sides by dark rims, especially on distal border near epigastric furrow (Fig. 250); deep brown, flask-shaped spermathecal bodies are visible in part, inside central orifice (Fig. 250). Thick, strongly coiled, whitish ducts of spermathecae lie above and along sides of sclerotized spermathecal bodies (Fig. 251).

Distribution: Israel, known only from the type locality, Husan, Judean Hills (11). An adult female was found in May.

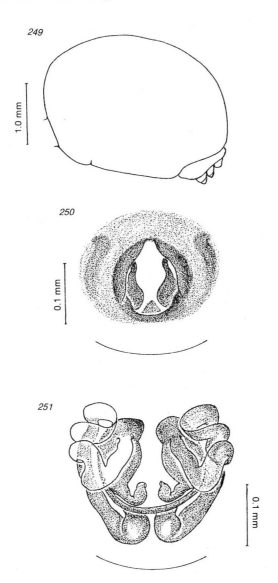

Figs 249–251: *Episinus fontinalis* Levy, 1985; female
249. opisthosoma, lateral view; 250. epigynum; 251. inner spermathecae, dorsal view

133

Genus ANELOSIMUS Simon, 1891

Annls Soc. entom. Fr. 60:11

Fig. 252

Type-species: *Theridium eximium* Keyserling, 1884.

Medium-sized theridiids about 2 to 6 mm in length (Fig. 252). Carapace longer than wide, without modifications in eye or clypeal regions. Eyes about equal in size, lateral eyes of anterior and posterior rows touching. Posterior tip of sternum projecting between coxae of fourth legs. Chelicerae with promarginal and retromarginal teeth. First pair of legs longest, third pair shortest. Opisthosoma usually oval, longer than wide or high, often with median, dorsal markings. Minute colulus bears two setae or only the two setae present between anterior spinnerets. Female usually with only one pair of internal spermathecae. Palpus of male with full complement of sclerites.

Anelosimus spiders resemble *Theridion* species but can be separated by the presence of retromarginal teeth on the chelicerae and by the colulus. About three dozen species are known so far from throughout the world; species of *Anelosimus* are abundant in Chile in particular. Many old types not yet re-examined, however, and the little known Asiatic tropical fauna may house many more. Some American species build communal webs displaying various degrees of social organization. This phenomenon is not known in the two *Anelosimus* species of Israel.

Key to the Species of *Anelosimus* in Israel

1. Opisthosoma, on dorsum, marked with distinct median band formed by successive, joined black spots (Fig. 254)   **A. aulicus** (C.L. Koch)
- Opisthosoma with indistinct pattern on back   **A. giladensis** Levy & Amitai

252

0.5 mm

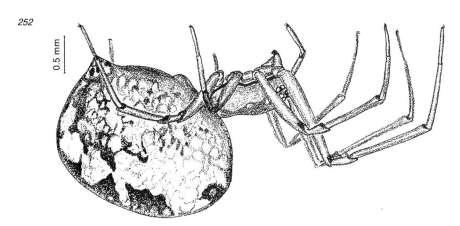

Fig. 252: *Anelosimus aulicus* (C.L. Koch, 1838); female,
lateral view of spider

134

Anelosimus aulicus (C.L. Koch, 1838)
Figs 252–259

Theridium aulicum C.L. Koch, 1838, *Die Arachniden*, 4:115, pl. 323; Roewer, 1942, *Katalog der Araneae*, 1:461; Bonnet, 1959, *Bibliographia Araneorum*, 2(5):4449.
Anelosimus aulicus —. Levi, 1956, *Trans. Am. microsc. Soc.* 75:412; Levy & Amitai, 1982a, *J. Zool. Lond.* 196:124.

Length of male 2.4–3.0 mm, female 2.8–4.2 mm. Coloration of carapace yellow-brown with deep brown to black, broad mid-dorsal band, widening anteriorly beyond posterior row of eyes; margins encircled with black (Fig. 253). Sternum yellow-brown in middle to dark brown on sides and black margins. Legs light brown. Opisthosoma on dorsum with large, conspicuous black spots connected by slender stalks, extending to spinnerets (Fig. 254); sides on upper parts brown mottled with white, and on lower

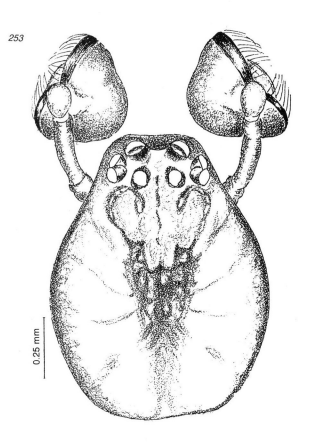

253

0.25 mm

Fig. 253: *Anelosimus aulicus* (C.L. Koch, 1838); male,
carapace and palpi, dorsal view

parts mottled with black markings (Fig. 252); venter grey to dark brown along middle, with elongated , light blotches on sides, behind epigastric furrow; brown spinnerets surrounded by six conspicuous, black spots (Fig. 255). Only two fine setae are present between anterior spinnerets.

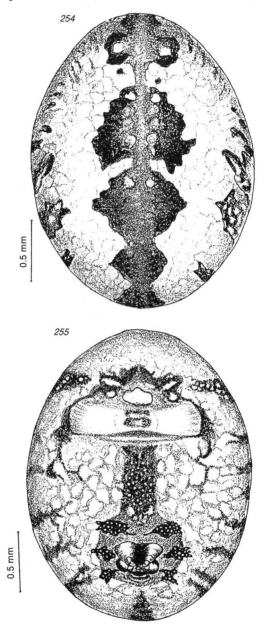

Figs 254–255: *Anelosimus aulicus* (C.L. Koch, 1838); female, opisthosoma
254. dorsal surface; 255. ventral surface

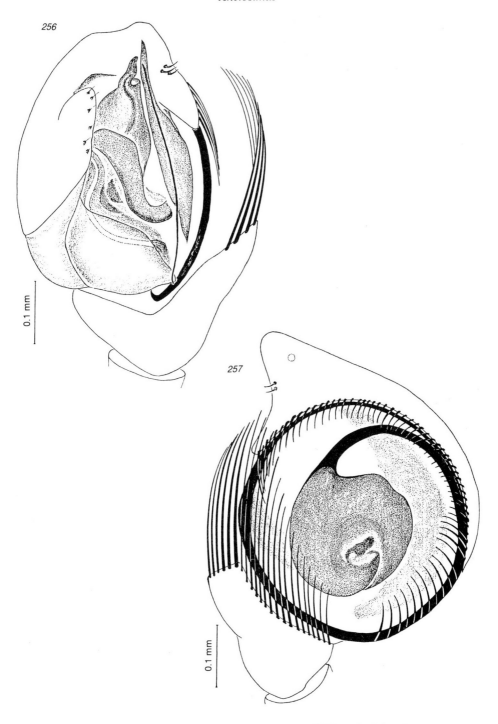

Figs 256–257: *Anelosimus aulicus* (C.L. Koch, 1838); male, left palpus
256. mesal view; 257. retrolateral view

Male Chelicera: Promargin with 3 teeth, retromargin with 3–4 denticles.

Male Palpus: Apical joint is rotated sideways and the cone-shaped back of the cymbium is, thus, turned mesally (Fig. 253). Tibia, along upper edges, bearing long, thick bristles, covering greater parts of retrolateral side of bulb (Figs 256, 257); additional, shorter bristles visible around outer margins of cymbium. Filiform portion of embolus appressed to edges of rounded ectal disc of bulb, encircling bulb about four or more times, ending on distal mesal side (Figs 256, 257).

Female Chelicera: As in male.

Female Epigynum: Almost indistinct on the outside, with only a narrow slit visible above light, membranous fold (Fig. 258). Orifices of internal ducts, hardly visible, even in dissection, placed on far ends of slit (Fig. 259); ducts appressed to small, dark spermathecal bodies in many tight coils (fourteen counted in one case; Fig. 259).

Distribution: Throughout southern Europe, Malta, Madeira, Canary Islands, Isle of St. Helena, northern Africa and the Middle East.

Israel: Throughout the country.

Adults of both sexes are found on plants from February to October. A female was taken in Jerusalem, in May, with a tough coated, oval egg sac of about 3 mm in diameter. The sac was filled with threads surrounding about forty eggs; diameter of an egg is about 0.5–0.6 mm.

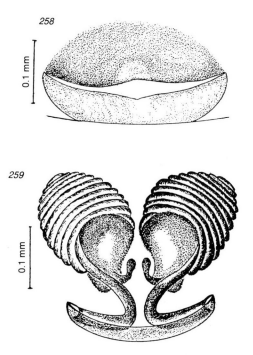

Figs 258–259: *Anelosimus aulicus* (C.L. Koch, 1838); female
258. epigynum; 259. inner spermathecae, dorsal view

138

Anelosimus giladensis Levy & Amitai, 1982
Figs 260–264

Anelosimus giladensis Levy & Amitai, 1982a, *J. Zool. Lond.* 196:127.

Length of male 2.9–3.2, female 3.0–5.7 mm. Carapace light brown with dorsal, dark rays. Male with stridulatory ridges posteriorly, above pedicel. Sternum uniformly yellow-brown. Legs yellowish to light brown without markings. Opisthosoma grey, pattern indistinct; male with brown, narrow band in front above pedicel. Colulus distinct, bearing two setae.

Male Chelicera: Promargin with three distinct teeth, retromargin with one denticle.
Male Palpus: Swollen basal division of embolus placed on ectal side of bulb (Figs 260–262). Filiform portion of embolus rising as fine, black stylet, turning distally on to cymbium, encircling ectal side of bulb down under raised edges of tibia and appearing thickly ensheathed on mesal side (Figs 260–262); conductor with raised, triangular portion and pointed tip (Figs 260–262); paracymbial-hook visible inside, on back of cymbium (Fig. 261).

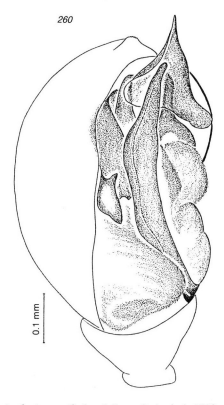

260

0.1 mm

Fig. 260: *Anelosimus giladensis* Levy & Amitai, 1982; male, left palpus,
mesal view

139

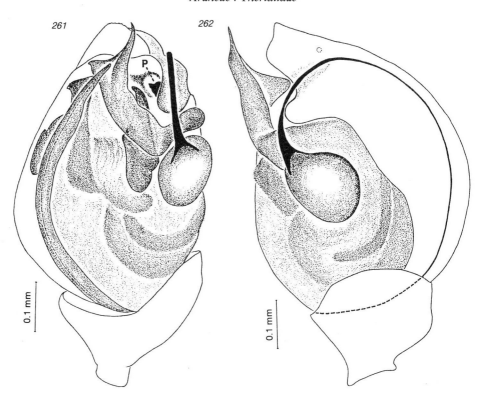

Figs 261–262: *Anelosimus giladensis* Levy & Amitai, 1982; male, left palpus
261. ventral view; P – paracymbial-hook; 262. retrolateral view

Female Chelicera: As in male.

Female Epigynum: Plate protrudes with fleshy, almost transparent, lip-like extension over edges of epigastric furrow (Fig. 263); two crescentic orifices visible on extended epigynal lip and dark silhouettes of internal organs discernible through integument (Fig. 263). Deep brown to black, obtuse spermathecal bodies are joined medially by a thick tube (Fig. 264); ribbon-like, tightly coiled ducts are placed obliquely between spermathecae and epigastric furrow (Fig. 264).

Distribution: Israel.

Israel: Upper Galilee (1) and near Lake Kinneret to the Carmel Ridge (3) and the Judean Hills (11).

Adult males are found in April, adult females from February to May. A female with two spherical egg sacs, 4.0 and 6.0 mm in diameter, each containing numerous eggs was found near Lake Kinneret in April.

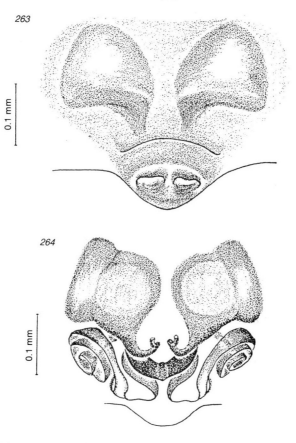

Figs 263–264: *Anelosimus giladensis* Levy & Amitai, 1982; female
263. epigynum; 264. inner spermathecae, dorsal view

Genus EURYOPIS Menge, 1868
Schr. naturf. Ges. Danzig (N.F.) 2:174.
Figs 265–270

Type-species: *Micryphantes flavomaculatus* C.L. Koch, 1836.

Medium-sized theridiids, total length about 1 to 6 mm, usually 2–3 mm. Prosoma often stout with a high clypeus in males, usually lower in females (Figs 265, 266, 275, 276, 284). Size of anterior median eyes variable relative to size of other eyes (Figs 265, 275). Anterior lateral and posterior lateral eyes touching or nearly touching. Median eye quadrangle usually wider in front than behind (Figs 267, 269). Chelicerae lacking teeth; fang long and thin (Fig. 268). Fourth pair of legs longest, length of remaining pairs varies slightly. Opisthosoma usually triangular, pointed behind (Fig. 270).

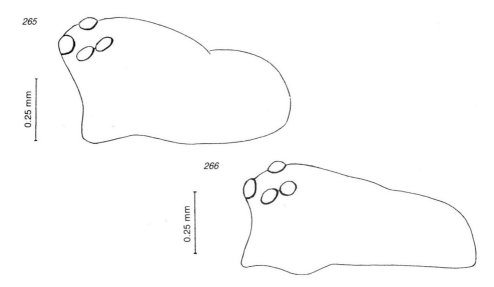

Figs 265–266: *Euryopis acuminata* (Lucas, 1846)
265. male, prosoma, lateral view; 266. female, prosoma, lateral view

Figs 267–268: *Euryopis acuminata* (Lucas, 1846); male
267. carapace, dorsal view of anterior part; 268. tip of left chelicera, inner view

Colulus usually visible as two small setae. Female usually has two pairs of internal spermathecae.

Euryopis spiders are not known to spin webs, but run on the ground and among vegetation, preying on ants. In some cases the spider and ant show remarkable similarity in colour and body shape. Sometimes, however, these differ entirely in size and form.

Euryopis species are distributed worldwide. A few dozen altogether, with many Old World species known only from their first description of only one sex. Considerable variation in colour, occasionally also in prosoma shape, is known at present in some species and many synonyms are thus disclosed, particularly when old types are studied along with fresh material. Out of nearly half a dozen *Euryopis* species formerly described from Israel only two are considered valid and a third was newly added (Levy & Amitai, 1981a).

142

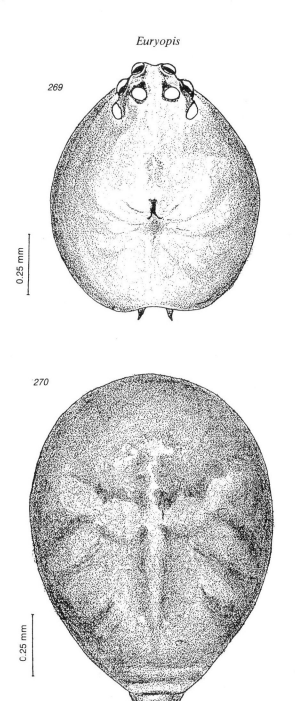

Euryopis

269

270

Fig. 269–270: *Euryopis acuminata* (Lucas, 1846); male
269. carapace, dorsal view; 270. opisthosoma, dorsal surface

143

*Key to the Species of **Euryopis** in Israel*

1. Opisthosoma brown to black, sometimes with an indistinct, light pattern on back (Fig. 270)
    E. acuminata (Lucas)
–   Opisthosoma black with a series of distinct white, silvery spots on back (Fig. 277)   2
2. Tip of conductor of male palpal organ accompanies tip of embolus, both extending to about the same height (Fig. 280); female with a distinct, sclerotized, shield-shaped epigynal plate (Fig. 282)   **E. sexalbomaculata** (Lucas)
–   Tip of conductor of male palpal organ recurves on itself, turning away from embolar tip (Fig. 285); female unknown   **E. hebraea** Levy & Amitai

Euryopis acuminata (Lucas, 1846)
Figs 265–274

Theridion acuminatum Lucas, 1846, *Explor. scient. Algér. Zool.* 1:268, pl. 17, fig. 10.
Euryopis acuminata —. Simon, 1873, *Mém. Soc. r. Sci.* Liège (2)5:117; Levy & Amitai, 1981a, *Bull. Br. arachnol. Soc.* 5(4):178.

Length of male 2.0–2.4 mm, female 2.5–3.5 mm. Coloration of prosoma black or yellow-brown, sometimes with a dark marginal line (Fig. 269). Legs yellowish-brown to dark grey, sometimes with light and dark parts giving a broadly banded appearance. Opisthosoma brownish to shiny black; sometimes slightly lighter but dull markings visible on back (Fig. 270).

Male Palpus: Rather peculiar among *Euryopis* species. Apically furnished with a large, sclerotized, partially bent and pointed prominence (Figs 271, 272); outside, at base of prominence, under apical edge of cymbium, rises a dark, obtuse and roughened protuberance (Fig. 272).

Female Epigynum: Plate with a large, central, partly rounded orifice from which a broad, shallow canal extends to epigastric furrow (Fig. 273); long tubes of internal organs sometimes discernible through integument. Internally, two long, straight, thick tubes run parallel, medially (Fig. 274); the tubes then form strong coils, partly surrounding each of the two pairs of spermathecae (Fig. 274).

Distribution: Southern Europe from Portugal to Bulgaria, Greece and Malta. Northern Africa, from Morocco to Egypt and Eritrea, and in Israel.

Israel: Throughout north and central parts of the country, along the Coast, in the mountainous areas and Plains of the Jordan (1–13, 18).

Adults of both sexes are found almost around the year, mainly from March to October. A female with a white, subspherical egg sac, about three mm in diameter fastened to a twig, was taken in September. The sac contained 22 eggs, each about 0.4 mm in diameter. *Euryopis acuminata* is often observed feeding on black *Tapinoma* ants. Actually these two, spider and ant, are hardly distinguishable in the field, but *E. acuminata* was also observed to feed on other ants.

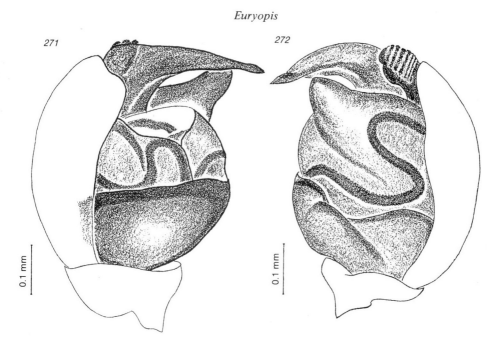

Figs 271–272: *Euryopis acuminata* (Lucas, 1846); male, left palpus
271. mesal view; 272. retrolateral view

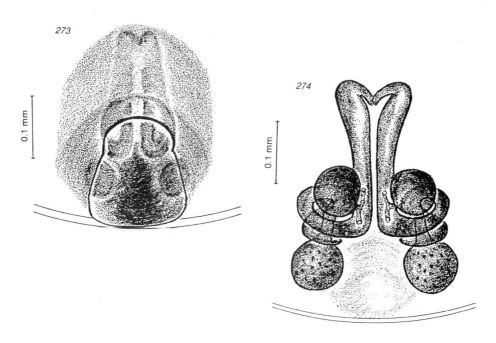

Figs 273–274: *Euryopis acuminata* (Lucas, 1846); female
273. epigynum; 274. inner spermathecae, dorsal view

Euryopis sexalbomaculata (Lucas, 1846)
Figs. 275–283

Theridion sexalbomaculatum Lucas, 1846, *Explor. scient. Algér. Zool.* 1:265, pl. 17, fig. 8.
Euryopis sexalbomaculata —. Simon, 1881, *Les Arachnides de France* 5(1):130; Levy & Amitai, 1981a, *Bull. Br. arachnol. Soc.* 5(4):180.

Length of male 2.7–2.9 mm, female 2.8–3.0 mm. Colour of prosoma deep brown, eye region black (Figs 275, 276). Legs yellow and partly black, mainly on the longer segments. Opisthosoma black with a series of distinct white, silvery spots on back, sides and venter (Figs 277–279); the spots may vary slightly in size, sometimes the hinder ones become almost confluent or connected by white streaks.

Fig. 275: *Euryopis sexalbomaculata* (Lucas, 1846); male,
prosoma, lateral view

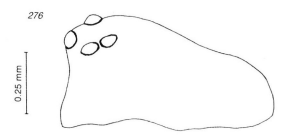

Fig. 276: *Euryopis sexalbomaculata* (Lucas, 1846); female,
prosoma, lateral view

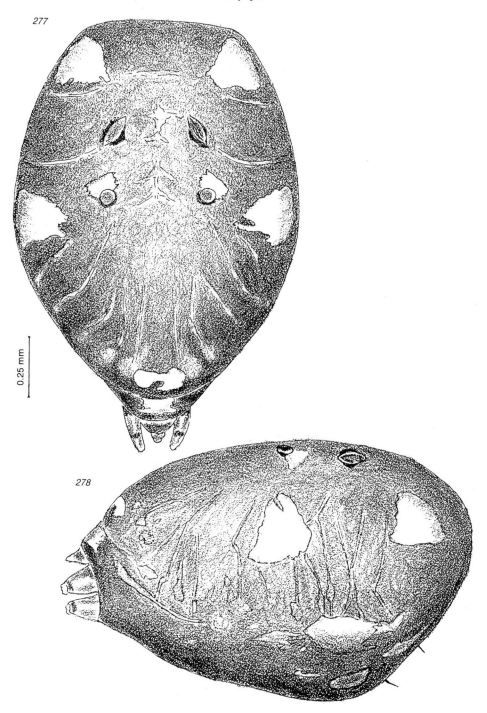

Figs 277–278: *Euryopis sexalbomaculata* (Lucas, 1846); male, opisthosoma
277. dorsal surface; 278. lateral view

279

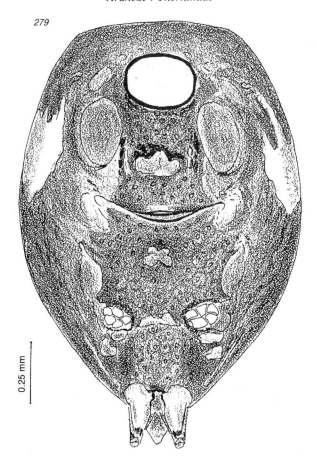

Fig. 279: *Euryopis sexalbomaculata* (Lucas, 1846); male, opisthosoma, ventral surface

Male Palpus: Cymbium rather broad apically and extended to a pointed tip projecting about as high as tips of embolus and conductor (Figs 280, 281); tip of embolus twists partly around tip of conductor, and both point upwards (Figs 280, 281).

Female Epigynum: Plate covered centrally by a dark brown shield (Fig. 282); an atrium extending below basal rim of shield is sometimes concealed by pieces of embedded brown material. Dark oval bodies of internal organs sometimes visible through integument on distal part of epigynum, close to epigastric furrow (Fig. 282). Two inner pairs of rounded spermathecae connected by broad, deep brown, sclerotized band (Fig. 283).

Distribution: Algeria, Tunisia, Libya, Greece, Israel.

Israel: Near Lake Kinneret, Lower Galilee (2) and Judean Hills (11).

Adult males and females are found under stones in April and May.

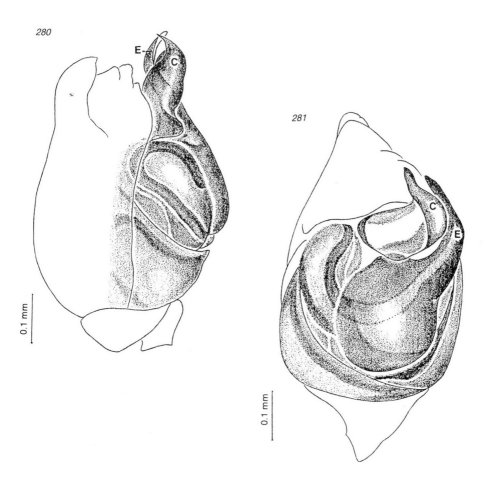

Figs 280–281: *Euryopis sexalbomaculata* (Lucas, 1846); male, left palpus
280. mesal view; 281. ventral view; C – conductor, E – embolus

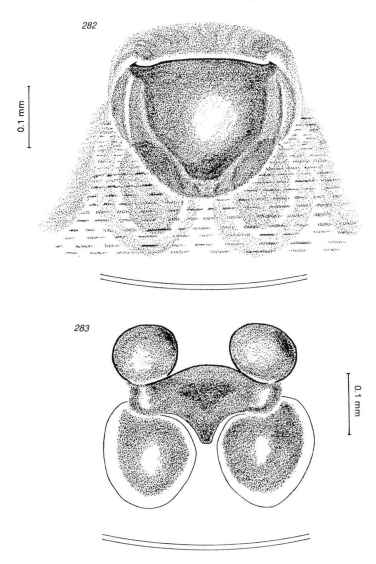

Figs 282–283: *Euryopis sexalbomaculata* (Lucas, 1846); female
282. epigynum; 283. inner spermathecae, dorsal view

Euryopis hebraea Levy & Amitai, 1981
Figs 284–286

Euryopis hebraea Levy & Amitai, 1981a, *Bull. Br. arachnol. Soc.* 5(4):182.

Length of male 2.7 mm, female unknown. Coloration of prosoma brown (Fig. 284).
Opisthosoma black with white spots arranged as in *E. sexalbomaculata.*

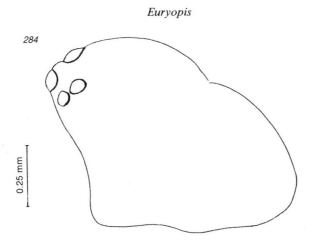

Fig. 284: *Euryopis hebraea* Levy & Amitai, 1981; male,
prosoma, lateral view

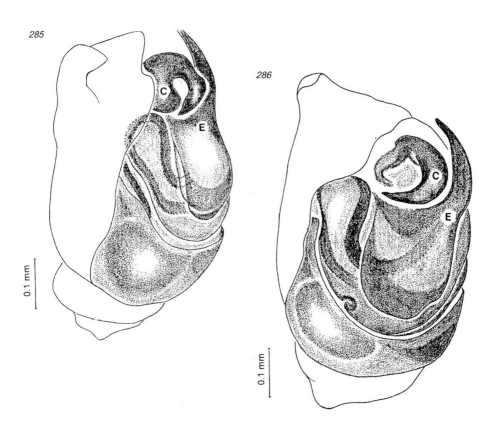

Figs 285–286: *Euryopis hebraea* Levy & Amitai, 1981; male, left palpus
285. mesal view; 286. ventral view; C – conductor, E – embolus

Male Palpus: Cymbium broad apically and extended to a pointed and slightly bent projection (Figs 285, 286); arched tip of embolus points upwards while tip of conductor recurves on itself, pointing inwards (Figs 285, 286).

Distribution: Israel: known only from the type locality, Jerusalem.

One adult male was found thus far in April.

Genus DIPOENA Thorell, 1869
Nova Acta R. Soc. Scient. Upsal. (3) 7:91.
Figs 287–289

Type-species: *Atea melanogaster* C.L. Koch, 1837.

Small spiders about 1.0 to 5.0 mm total length, many only about 2 mm (Fig. 287). Prosoma short, and in males, often very high (Figs 287, 294). Eye region usually projecting above concave clypeus. Anterior median eyes sometimes larger than others, and placed rather close to anterior lateral eyes (Figs 288, 297). Anterior lateral and posterior lateral eyes touching. Median eye quadrangle wider in front than behind (Figs 289, 296). Chelicerae small and without teeth (Fig. 288). First leg commonly the longest, third always the shortest. Opisthosoma usually almost spherical, wide above and higher than long (Figs 287, 298, 304). Colulus visible as two small setae.

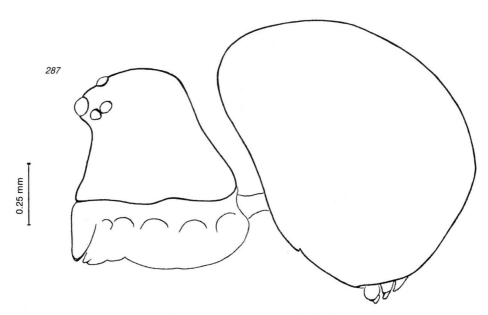

Fig. 287: *Dipoena convexa* (Blackwall, 1870); male,
lateral view of spider (legs omitted)

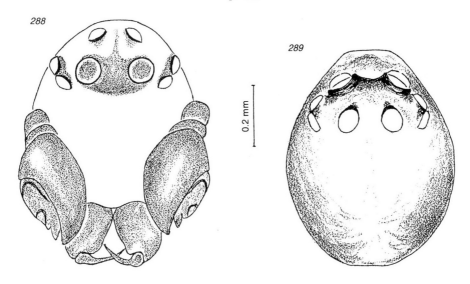

Figs 288–289: *Dipoena convexa* (Blackwall, 1870); male
288. carapace, chelicerae and palpi, frontal view; 289. carapace, dorsal view

Dipoena species are among the smallest theridiid spiders of Israel and are considered rare. They live in relatively moist places and were collected in Israel in dense vegetation and under stones. They may build small webs made of very fine threads and feed on ants, but these were not yet observed in Israel.

Numerous *Diopena* species have been described from many parts of the world but their occurrence in the Middle East was only recently established (Levy & Amitai, 1981a). *Dipoena* specimens are often scarce even in well-collected places in Europe and America, and many species are known from only one of the sexes. Three species are known in Israel.

Key to the Species of **Dipoena** in Israel

Males:

1.   Carapace about as high as wide, trapezoidal in profile (Fig. 287)

                                                 D. convexa (Blackwall)

–   Carapace distinctly wider than high   **D. braccata** (C.L. Koch)

Females:

1.   Opisthosoma with a distinct, coloured pattern on dorsal surface (Fig. 304)

                                                 D. galilaea Levy & Amitai

–   Opisthosoma uniformly black   2
2.   Carapace about as high as wide   **D. convexa** (Blackwall)
–   Carapace appreciably wider than high   **D. braccata** (C.L. Koch)

Dipoena convexa (Blackwall, 1870)
Figs 287–293

Theridion convexum Blackwall, 1870, *J. Linn. Soc. Lond.* 10:426.
Dipoena Convexa —. Simon, 1894. *Historie naturelle des Araignées* 1 (3):562; Roewer, 1942,
  Katalog der Araneae, 1:418; Bonnet, 1956, *Bibliographia Araneorum* 2(2): 1504; Brignoli,
  1967, *Fragm. Entomol.* (Roma) 4(10): 183.
Dipoena trapezoidalis Levy & Amitai, 1981a, *Bull. Br. arachnol. Soc.* 5(4):184. Vanuytven, Van
  Keer & Poot, 1994, *Nwsbr. Belg. Arachnol. Ver.* 9(1):2 (synonymy with *D. convexa*).

Length of male 1.7 mm, female not yet collected in Israel. Coloration of high trapezoidal carapace brown (Figs 287–289). Legs yellowish-brown with only the terminal
segments of the palpi brown. High opisthosoma, uniformly black (Fig. 287).
Male Palpus: Small cone-shaped radix resembles the median apophysis, the two being
hardly separated (Figs 290, 291). Apical rims of tegulum extend into a sclerotized,
raised lobe (Figs 290, 291).

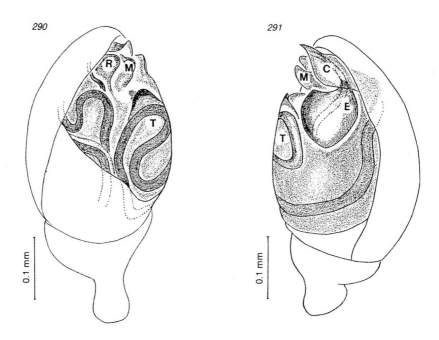

Figs 290–291: *Dipoena convexa* (Blackwall, 1870); male, left palpus
290. mesal view; 291. retrolateral view; C – conductor, E – embolus,
M – median apophysis, R – radix, T – tegulum

Female Epigynum: Drawings provided are of a female from France (Figs. 292, 293).
Distribution: Southern Europe to Balkan countries, Algeria, Tunisia, Israel.
Israel: On Carmel Ridge (3) and near Dorot, northern Negev (15).
A male on the Carmel was found under a stone in December, the one from the Negev was collected in a hole in the ground in February.

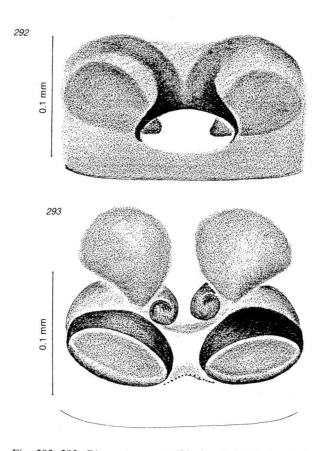

Figs 292–293: *Dipoena convexa* (Blackwall, 1870); female from France
292. epigynum; 293. inner spermathecae, dorsal view

Dipoena braccata (C.L. Koch, 1841)
Figs 294–302

Theridium braccatum C.L. Koch, 1841, *Die Arachniden*, 8:85.
Dipoena braccata —. Simon, 1894, *Histoire naturelle des Araignées* 1(3):562; Roewer, 1942, *Katalog der Araneae*, 1:418; Bonnet, 1956, *Bibliographia Araneorum* 2(2):1503; Levy & Amitai, 1981a, *Bull. Br. arachnol. Soc.* 5(4):185.

Length of male 1.2–1.4 mm, female 2.0–2.9 mm. Coloration of prosoma brown with black margins and dark eye region, or uniformly deep brown to black (Figs 294–297). Legs light brown. Opisthosoma uniformly black (Fig. 298).

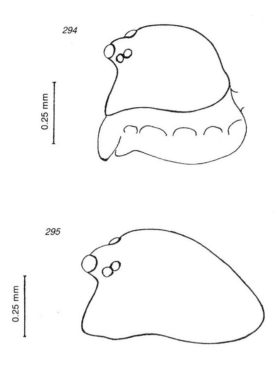

Figs 294–295: *Dipoena braccata* (C.L. Koch, 1841)
294. male, prosoma, lateral view (legs omitted);
295. female, prosoma, lateral view

Male Palpus: Small. Fine, slightly undulating tips of conductor and radix placed close to each other (Figs 299, 300); cone-shaped median apophysis differs distinctly in form from pointed radix (Figs 299, 300).

Female Epigynum: Small with details of external structure hardly visible: semicircular sclerotization supported on each side by a dark, arched band (Fig. 301). Internal spermathecae consist of two large compact bodies connected by short, straight ducts to a pair of small oval bodies (Fig. 302).

Distribution: Central and southern Europe, North Africa, Israel.

Israel: Mt Hermon (1650 m; 19), Coastal Plain (8) and Judean Hills (11).

Adult males were collected in March, May and August and adult females in May and July. Specimens were found on moist *Adiantum* ferns, as well as on the ground in grass and among pine needles.

156

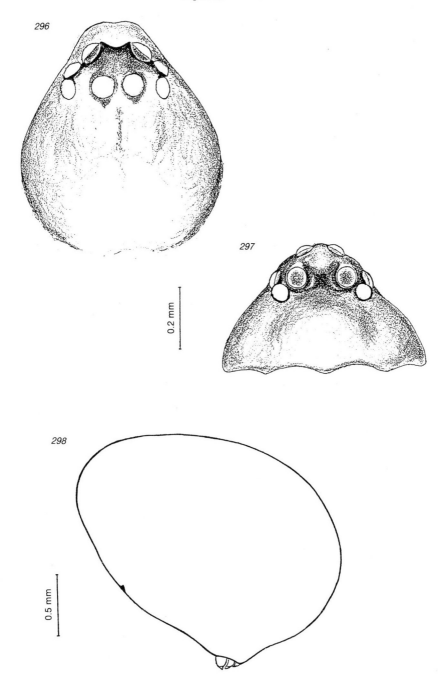

Figs 296–298: *Dipoena braccata* (C.L. Koch, 1841); female
296. carapace, dorsal view; 297. carapace, frontal view;
298. opisthosoma, lateral view

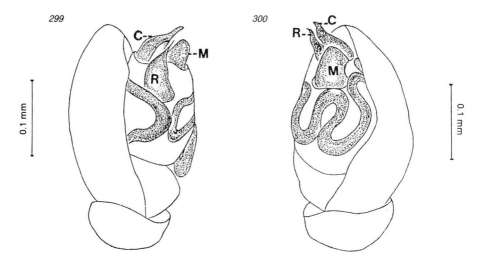

Figs 299–300: *Dipoena braccata* (C.L. Koch, 1841); male, left palpus
299. mesal view; 300. retrolateral view; C – conductor, M – median apophysis, R – radix

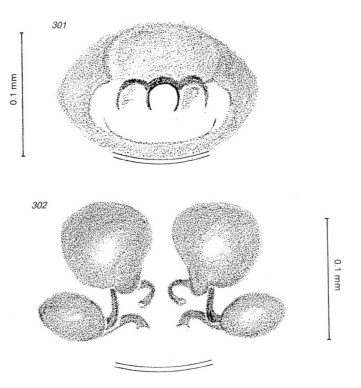

Figs 301–302: *Dipoena braccata* (C.L. Koch, 1841); female
301. epigynum; 302. inner spermathecae, dorsal view

158

Dipoena galilaea Levy & Amitai, 1981
Figs 303–307

Dipoena galilaea Levy & Amitai, 1981a, *Bull. Br. arachnol. Soc.* 5(4):186.

Male unknown. Length of female 1.75–2.10 mm. Carapace brown, mottled with black markings (Fig. 303). Legs yellow with slightly dark distal articulations. Opisthosoma black with a distinct white pattern dorsally (Figs 304, 305).

Fig. 303: *Dipoena galilaea* Levy & Amitai, 1981; female,
prosoma, lateral view

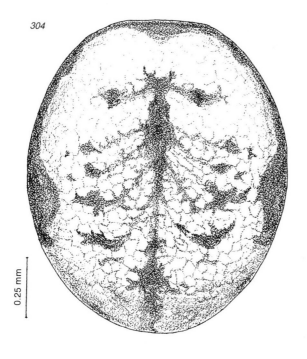

Fig. 304: *Dipoena galilaea* Levy & Amitai, 1981; female,
opisthosoma, dorsal surface

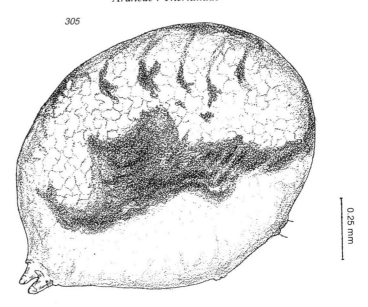

Fig. 305: *Dipoena galilaea* Levy & Amitai, 1981; female,
opisthosoma, lateral surface

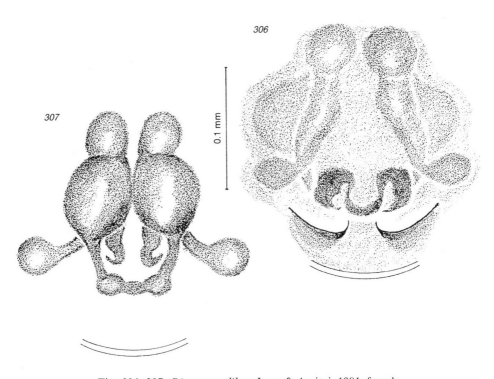

Figs 306–307: *Dipoena galilaea* Levy & Amitai, 1981; female
306. epigynum; 307. inner spermathecae, dorsal view

160

Female Epigynum: Plate with two, fine sclerotized shallow cups, placed close to epigastric furrow (Fig. 306). Spermathecal orifices, above cups, partly surrounded by dark, winding band (Fig. 306). Internal organs partly visible through integument. Of the two pairs of spermathecal bodies, one pair is rather large, elongated and partly constricted apically (Fig. 307).

Distribution: Israel, northern Galilee (1).

Adult females were found in August in moist places with dense vegetation. One female was taken with an egg sac containing five oval eggs, each about 0.6–0.7 mm in diameter.

Genus COSCINIDA Simon, 1894
Histoire naturelle des Araignées 1(3):529.
Figs 308–311

Type-species: *Coscinida tibialis* Simon, 1895.

Small theridiid spiders of usually less than 2 mm total length. Carapace slightly longer than wide (Fig. 308). Eyes relatively large and closely grouped (Figs 308, 310); ocular area black. Clypeus with a distinct recess below eye region (Fig. 309). Median eyes

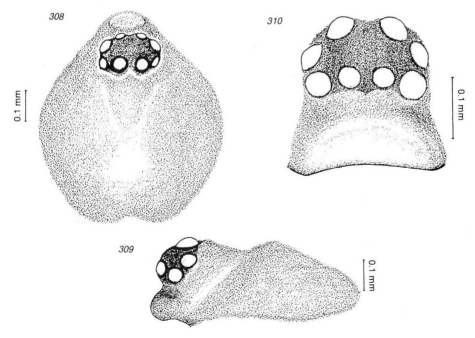

Figs 308–310: *Coscinida tibialis* Simon, 1895; male
308. carapace, dorsal view; 309. prosoma, lateral view; 310. carapace, frontal view

161

closer to laterals than to each other; anterior-lateral eyes and posterior-laterals almost touching. Sternum posteriorly rounded. Chelicerae with a small, mesal, anterior boss; no retromarginal teeth. Fourth pair of legs longest, third shortest. Opisthosoma oval, longer than wide (Fig. 311). Colulus absent. Palpus of male with all sclerites; upper-mesal or distal margin of cymbium drawn into an external, pointed hook (not a paracymbial-hook; Fig. 312).

The closely spaced large eyes and, in particular, the external hook on the male palpus are characteristic features of *Coscinida*. Only very few species, all from the Old World are included in *Coscinida* thus far: five from Africa and six from south-east Asia. The species recently found in Israel represents the only record of this genus from the Middle East (Levy, 1985a).

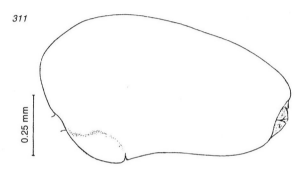

Fig. 311: *Coscinida tibialis* Simon, 1895; male,
opisthosoma, lateral view

Coscinida tibialis Simon, 1895
Figs 308–315

Coscinida tibialis Simon, 1895, *Annls Soc. ent. Fr.* 64:137; Roewer, 1942, *Katalog der Araneae*, 1:447; Bonnet, 1956, *Bibliographia Araneorum*, 2(2):1240; Levy, 1985a, *J. Zool. Lond.* 207:112.

Length of male 1.75–1.80 mm, female not yet collected in Israel; males and females from Algerian type material are of the same size as Israeli males. Coloration of carapace light brown with edges marked by a fine, black line; eye region black (Figs 308–310); no stridulating ridge posteriorly in male. Light brown, shield-like sternum, slightly convex; border with labium barely visible. Legs yellow. Brown oval opisthosoma (Fig. 311) dorsally flecked with a few irregular, greyish spots; venter dark along central portion.

Male Palpus: Tibia small. Mesal margins of cymbium distally armed with a black sclerotized, pointed hook (Figs 312, 313). Light brown median apophysis consists of a broad, fleshy basal portion and a whitish, slender protuberance rising from inside and extending to tip of bulb (Figs 312, 313); embolus rises on retrolateral side of bulb and extends into a central deep groove formed by folds of conductor (Figs 313, 314); walls of conductor groove attenuate distally forming two black, pointed processes (Fig. 312).

Female Epigynum: Drawings provided are of female from type material from Algeria (Fig. 315).

Distribution: Algeria, Angola, Israel.

Israel: Regba in northern Coastal Plain (4).

Two males were found on the ground. Biological information on these tiny spiders is completely wanting.

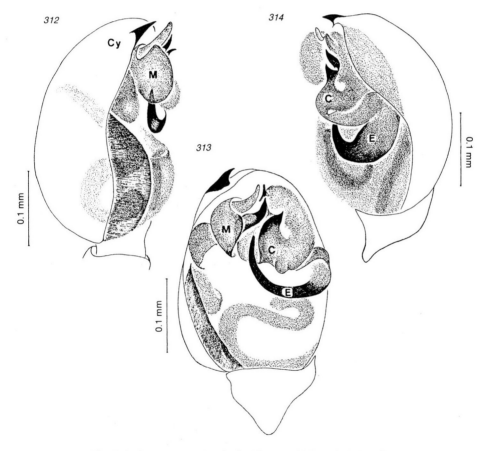

Figs 312–314: *Coscinida tibialis* Simon, 1895; male, left palpus
312. mesal view; 313. ventral view; 314. retrolateral view; C – conductor, Cy – cymbium, E – embolus, M – median apophysis

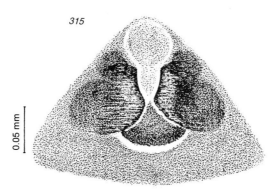

Fig. 315: *Coscinida tibialis* Simon, 1895; female from Algeria, epigynum

Genus ACHAEARANEA Strand, 1929
Latv. Univ. Rak. 20:11
Fig. 316

Type-species: *Argyrodes trapezoidalis* Taczanowski, 1873.

Theridiids ranging between 1 and 6 mm in body length or a little more. Carapace slightly longer than wide and without modifications in eye or clypeal region. Eyes

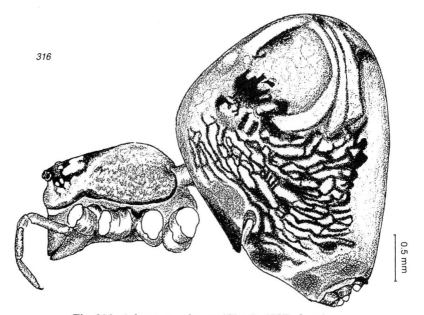

Fig. 316: *Achaearanea lunata* (Clerck, 1757); female,
lateral view of spider (legs omitted)

about equal in size, laterals touching. Posterior-median eyes sometimes slightly oval. Median quadrangle of eyes about square. Sternum posteriorly truncate. Chelicerae without teeth or, with mesal, promarginal small tooth or boss; no retromarginal teeth. First pair of legs longest, third pair shortest. Opisthosoma usually (but not always) higher than long, often with a distinct pattern of streaks on sides (Fig. 316). Colulus absent. Female with one pair of internal spermathecae. Palpus of male without median apophysis; radix and embolus may form one continuous sclerite.

Achaearanea spiders closely resemble *Theridion* species but can be distinguished by the high opisthosoma with the streaky pattern on sides and the relatively simple palp, lacking the median apophysis. Nearly one hundred species were described so far mainly from the Neotropics. *Achaearanea* is represented in Israel by only one species.

Achaearanea lunata (Clerck, 1757)
Figs 316–321

Araneus lunatus Clerck, 1757, *Aranei Suecici*:52.
Theridion lunatum —. Latreille, 1819, *Nouv. Dictionnaire d'Hist. Nat. Article* 34:12; Roewer, 1942, *Katalog der Araneae*, 1:474; Bonnet, 1959, *Bibliographia Araneorum*, 2(5):4485.
Achaearanea lunata —. Levi, 1955, *Am. Mus. Novit.* No. 1718:8; Levy & Amitai, 1982a, *J. Zool. Lond.* 196:121.

Length of male 2.7 mm, female 3.1–4.9 mm. Coloration of carapace red-brown with black, mid-dorsal and marginal markings. Sternum brown with black margins, labium broadly attached with no separating line. Legs yellow to light brown with dark markings close to articulations, most distinct on tibiae of fourth legs. Opisthosoma reddish-brown on back with large, light creamy patches in front and white stripes arching down sides; sides black with dense, bright white markings (Fig. 316); venter with white, conspicuous spot behind epigastric furrow, and light patches in front of spinnerets, but light brown spinnerets surrounded by a deep brown marking.

Male Palpus: Thick, large embolus together with radix form one continuous sclerite that is partly covered in the middle by a fine membranous fold (Figs 317–319); distal portion of embolus slightly winding. Large, fleshy conductor distinctly projects banner-like above cymbium (Fig. 317); external surface of conductor covered with small, dark tubercles (Figs 317–319).

Female Epigynum: Plate with a central depression bordering epigastric furrow; depression surrounded by dark brown, thick raised band, coiling inwards on anterior margin (Fig. 320). Spermathecae with yellow-brown roughened surface, attaching medially to very thick, straight, black tubes (Fig. 321).

Distribution. Throughout Europe to Siberia, China and Japan. Northern Israel, presumably in Lebanon.

Israel: Along northern borders of Upper Galilee (1).

All specimens in Israel were found on tree stems, in April and in August to November. A female with young emerging from an egg sac was taken in mid-October. The egg sac is brownish, spherical, tough coated, about 3.5 mm in diameter.

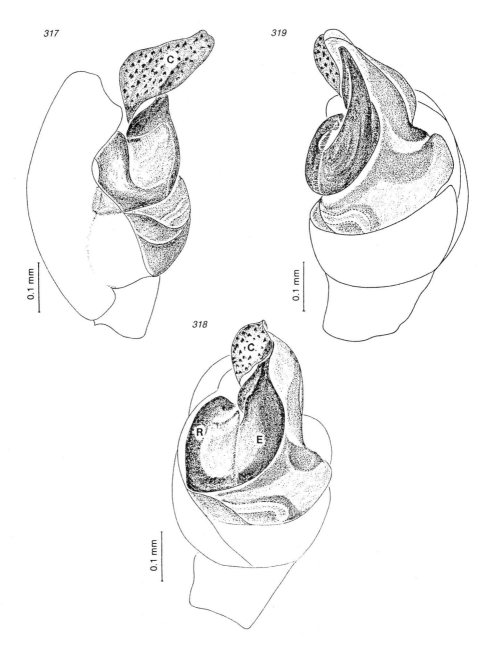

Figs 317–319: *Achaearanea lunata* (Clerck, 1757); male, left palpus
317. mesal view; 318. ventral view; 319. retrolateral view; C – conductor, E – embolus, R – radix

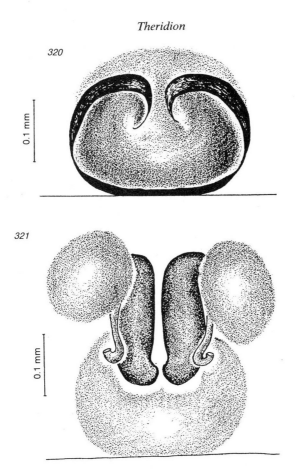

Figs 320–321: *Achaearanea lunata* (Clerck, 1757); female
320. epigynum; 321. inner spermathecae, dorsal view

Genus THERIDION Walckenaer, 1805
Tableau des Aranéides, Paris: 72.

Type-species: *Aranea picta* Walckenaer, 1802.

Small to medium-sized theridiid spiders, 1 to 7 mm total length. Carapace usually slightly longer than wide, without modification in eye or clypeus regions. Anterior eye row almost straight from in front, posterior row straight from above. Eyes about equal in size, lateral eyes touching. Posterior tip of sternum usually projecting bluntly between coxae of fourth legs. Chelicerae with no teeth or, with a mesal, anterior boss to two promarginal teeth; no retromarginal teeth. Legs moderate to long, first pair longest, third shortest. Opisthosoma more or less spherical. Colulus absent. Female with one pair of internal spermathecae. Palpus of male with all sclerites.

Theridion spiders are usually found hanging upside down in an irregular web suspended on plants or hidden in rock crevices or fissures in the soil. Many use very fine threads, often hardly noticeable unless they occassionally glisten in the sunlight or are covered with dust. *Theridion*, the largest genus of the theridiids, also considered as one of the largest genera amongst all spiders, embraces several hundred species. Many species have been transferred from *Theridion* to other genera, and still *Theridion* is most diverse, but further separation of small species groups seems impracticable. It is already difficult to recognize unaccompanied females and place them accurately in *Theridion* or *Achaearanea*.

Theridion species are found throughout the world, being most abundant in the warmer areas and the tropics. Probably only a portion of the species belonging to *Theridion* are known thus far. This is true also for the theridiid fauna of Israel that has not been exhausted, especially with *Theridion* in which additional new species should be expected. At present there are 21 species in Israel.

*Key to the Species of **Theridion** in Israel*

Males:

1.   Opisthosoma covered with small black dots (Figs 349, 358)   2
–   Opisthosoma otherwise coloured or pattern indistinct   3
2.   Light sternum surrounded by a black, scalloped band, and in the middle with a distinct, black spot (Fig. 357); palpus as in Figs 359–361; body length 3 mm or more
    T. nigropunctatum Lucas
–   Yellow sternum encircled only by a fine black line; palpus as in Figs 350–352; body length less than 2 mm   **T. accoense** Levy
3.   Opisthosoma on dorsum with anterior humps or with a posterior tubercle   4
–   Opisthosoma spherical or almost round, without tubercles on back   5
4.   Opisthosoma with a conspicuous protuberance (Fig. 401); palpus kidney-bean-shaped and very large (Fig. 401)   **T. uncinatum** Lucas
–   Opisthosoma in front with low, sideways projecting cones, sometimes indistinct (Fig. 391); palpus rotated outwards with embolar duct encircling bulb on ectal side (Fig. 390)
    T. dromedarius Simon
5.   Chelicerae with two promarginal denticles; palpus with pointed, hook-like apophysis (H) at base of conductor (C) and toothed projection (BP) branching off mesal embolus (Figs 322, 323)   **T. rufipes** Lucas
–   Chelicerae without teeth or with only a minute, mesal boss; palpus otherwise   6
6.   Palpus with subtegulum (ST) filling about proximal half of bulb (Figs 327, 340, 344); filiform portion of embolus short   7
–   Palpus with subtegulum filling less than proximal one third of bulb; filiform portion of embolus encircling major parts of bulb   10
7.   Palpus with large, distinct radix (R) visible from nearly all angles (Figs 340–342)
    T. vespertinum Levy
–   Palpus with short radix visible almost only from mesal side   8

8. Filiform portion of embolus hidden by leaf-like, membranous conductor and accompanying sclerite (Fig. 344)   **T. simile** C.L. Koch
-   Arched, filiform portion of embolus visible throughout; conductor partly sclerotized, not leaf-like   9
9. Conductor (C) with round, obtuse, low protuberance on ectal side (Fig. 328); median apophysis (M) not extending above conductor in mesal or ventral view (Figs 328, 329)   **T. melanurum** Hahn
-   Conductor with large, horn-like projection on ectal side; tip of median apophysis visible above conductor in mesal and ventral view (Figs 332, 333) **T. ochreolum** Levy & Amitai
10. Filiform portion of embolus rising distally from embolar basal division (Fig. 365); radix (R) pointed, finger-like, protruding at middle of mesal side; median apophysis (M) almost entirely hidden by circular extension of conductor (Fig. 364)   **T. melanostictum** O.P.-Cambridge
-   Filiform portion of embolus rising from ectal or proximal side of basal embolar division (Figs 371, 376, 382); radix otherwise protruding; median apophysis readily visible on mesal side or not visible at all as a distinct projecting sclerite   11
11. Large radix bulging along mesal margin of bulb, its proximal portion pointed and distally directed; median apophysis not visible (Figs 381, 382)   **T. negebense** Levy & Amitai
-   Radix otherwise; median apophysis visible as a distinct sclerite   12
12. Radix strongly protruding on proximal portion of bulb along edges of subtegulum (Fig. 375); median apophysis visible distally, above conductor, in ventral view, and most conspicuous on retrolateral side of palpus (Figs 376, 377)   **T. musivum** Simon
-   Radix lobe-like, slightly projecting at about middle of mesal side of bulb (Fig. 370); median apophysis not visible on retrolateral side of palpus (Fig. 372)   **T. hierichonticum** Levy & Amitai

Females:

1. Opisthosoma covered with small black dots (Fig. 358); light sternum surrounded by black, strongly scalloped band and in middle with a distinct black spot   **T. nigropunctatum** Lucas
-   Opisthosoma and sternum otherwise coloured or pattern indistinct   2
2. Opisthosoma on dorsum with anterior humps or, posteriorly, with a low bulge or a coned tubercle   3
-   Opisthosoma spherical or nearly round without modifications   5
3. Opisthosoma on dorsum with anterior, sideways projecting, coned humps (Fig. 391)   **T. dromedarius** Simon
-   Opisthosoma with only a posterior protuberance   4
4. Opisthosoma posteriorly with a conspicuous, coned tubercle (Fig. 402); dorsum with eight white spots arranged in two rows; epigynal plate relatively large and raised (Fig. 406)   **T. uncinatum** Lucas
-   Opisthosoma posteriorly with a low protuberance, sometimes visible only in profile (Fig. 408); dorsum with four black, distinct spots, sometimes with additional median ones; epigynum relatively small with central depression surrounded by semicircular, dark, raised rims (Fig. 409)   **T. pustuliferum** Levy & Amitai
5. Chelicerae with two promarginal denticles; epigynum usually hidden by a bulge of hardened secretion, covering a short V-shaped slit (Fig. 325)   **T. rufipes** Lucas

169

- Chelicerae without teeth or with only a minute, mesal boss; epigynum otherwise   6
6. Epigynum with a single, central depression containing two readily apparent circular openings of ducts   7
- Epigynum otherwise; if with central depression, duct openings not readily apparent   9
7. Epigynum placed on top of high, cone-like protuberance; openings of ducts filling entire epigynum (Fig. 373)   **T. jordanense** Levy & Amitai
- Epigynal plate not raised; openings of ducts not taking entire space of common depression   8
8. Large, semicircular, pit-like openings separated by dark, inward-tapering projection; central depression of epigynum with fine margins (Fig. 397)   **T. dafnense** Levy & Amitai
- Small, dark adjacent openings placed near centre, ducts near openings partly discernible; central depression with sclerotized rims on sides and on borders of epigastric furrow (Fig. 378)   **T. musivum** Simon
9. Epigynum with indistinct central depression   10
- Edges of central depression at least partly raised   11
10. Openings of ducts, bordering epigastric furrow, placed slightly apart or joined by a reddish bar (Figs 367, 368); spermathecae elongated and ducts partly coiled in tight helices (Fig. 369)   **T. melanostictum** O.P.-Cambridge
- Black-encircled openings of ducts placed at middle of epigynal plate far from epigastric furrow (Fig. 355); shape of spermathecae and arrangement of ducts otherwise (Fig. 356)   **T. hermonense** Levy
11. Epigynal plate folded, markedly raised, along borders of epigastric furrow, central depression, very small, circular or slightly oval, placed several times its diameter distant from epigastric furrow (Fig. 353)   **T. hemerobius** Simon
- Epigynal plate not raised on posterior borders; central depression of different shape and different distance from furrow   12
12. Epigynum with relatively deep central depression, entirely or partly surrounded by thick, sclerotized margins   13
- Central depression shallow, with sometimes hardly visible, fine, sclerotic rims   15
13. Central depression partly surrounded by funnel-like, thick semicircular rims (Fig. 399); spermathecae slightly elongated with ducts forming a much widened portion (Fig. 400)   **T. vallisalinarum** Levy & Amitai
- Central depression almost or entirely encircled by heavy rims; spermathecae and ducts otherwise   14
14. Epigynum with median, wide, slightly raised septum on bottom of central depression; openings of ducts slit shaped, visible centrally on sides of septum (Fig. 388); spermathecae partly surrounded by winding coils of ducts (Fig. 389)   **T. gekkonicum** Levy & Amitai
- Central depression entirely surrounded by thick rims; no septum; openings of ducts partly hidden by posterior margins of depression (Fig. 384); loops of coiled ducts not extending above or beyond spermathecae (Figs. 385–387)   **T. negebense** Levy & Amitai
15. Epigynal depression partly covered anteriorly by a transparent hood (Fig. 338)   **T. agaricographum** Levy & Amitai
- Epigynal depression without a hood   16
16. Central depression placed at a distance from epigastric furrow; rims on posterior border slightly funnel-like, raised (Fig. 347)   **T. simile** C.L. Koch

- Central depression of epigynum extending almost to borders of epigastric furrow; rims on sides of depression almost indistinct (Figs 330, 335)   17
17. Fine brown ducts of spermathecae forming two or more loops (Fig. 331)

T. melanurum Hahn
- Relatively thick brown ducts of spermathecae forming a single loop (Fig. 336)

T. ochreolum Levy & Amitai

Theridion rufipes Lucas, 1846
Figs 322–326

Theridion rufipes Lucas, 1846, *Explor. scient. Algér. Zool.* 1:263, pl. 16, fig. 5; Roewer, 1942, *Katalog der Araneae*, 1:459; Levi, 1957, *Bull. Am. Mus. nat. Hist.* 112:56; Bonnet, 1959, *Bibliographia Araneorum*, 2 (5):4522; Levy & Amitai, 1982a, *J. Zool. Lond.* 196:86.

Length of male 2.8–4.2 mm, female 3.7–6.5 mm. Coloration of carapace and legs yellowish-brown. Opisthosoma light brown to dark grey, occasionally with white spotted stripes, descending posteriorly. Posterior median eyes distinctly elongated, not round. Male with stridulating ridges on carapace, above pedicel.
Male Chelicera: Promargin with 2 teeth.
Male Palpus: Relatively large. Embolar division accommodated on mesal part of bulb (Figs 322, 323). Short, boss-like protuberance (BP) projecting at base of distal, tapering portion of embolus (Figs 322, 323); dark, sclerotized, hook-like, pointed apophysis (H) projecting from base of membranous conductor (Figs 322–324); large folds of conductor with roughened surface on retrolateral side (Figs 323, 324).
Female Chelicera: As in male.
Female Epigynum: Red, hard resinous matter, bulging in profile, usually covering epigynal plate. Opening, under resinous plug, in form of short, V-shaped slit placed at great distance from epigastric furrow (Fig. 325). Spermathecae very large, pear-shaped with thick, relatively short tubes (Fig. 326).
Distribution: Throughout warmer parts of the world.
Israel: Found primarily inside animal-breeding facilities, feeding on houseflies. Probably introduced into this fauna being usually found in association with man.

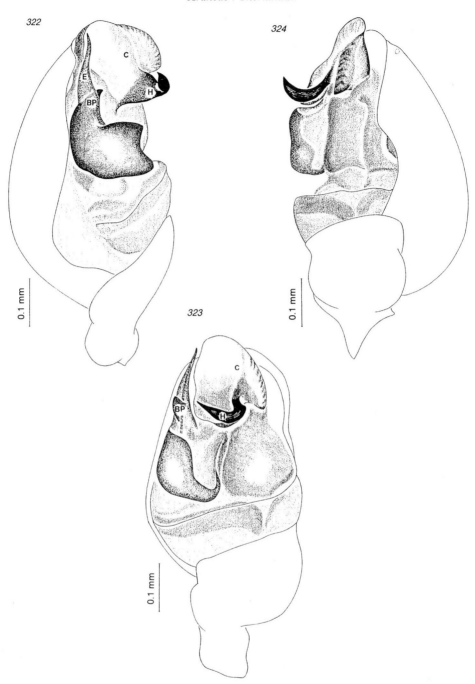

Figs 322–324: *Theridion rufipes* Lucas, 1846; male, left palpus
322. mesal view; 323. ventral view; 324. retrolateral view;
BP – boss-like protuberance, C – conductor, E – embolus, H – hook-like apophysis

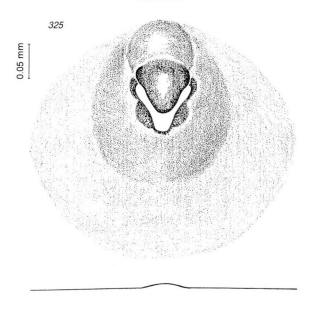

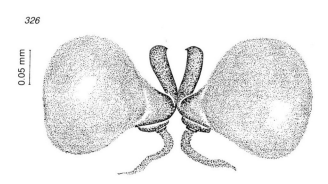

Figs 325–326: *Theridion rufipes* Lucas, 1846; female
325. epigynum; 326. inner spermathecae, dorsal view

Theridion melanurum Hahn, 1831

Figs 327–331

Theridion melanurum Hahn, 1831, *Monographie der Spinnen*, 6: pl. 3 fig. a, Nürnberg; Roewer, 1942, *Katalog der Araneae* 1:466; Levi, 1957, *Bull. Am. Mus. nat. Hist.* 112:55; Levy & Amitai, 1982a, *J. Zool. Lond.* 196:89.

Length of male 3.1–3.9 mm, female 3.3–4.5 mm. Carapace light to dark brown with an almost black, mid-dorsal mark. Sternum deep brown, darker along margins. Legs light to brown with dark patches or reduced annulations. Opisthosoma light brown to almost black with light, reddish-white, mid-dorsal, dentated band, most distinct on posterior part; venter with conspicuous yellow triangular spot, behind epigastric furrow; anterior portion of venter of male dark and markedly swollen.

Male Palpus: Relatively large. Subtegulum (ST) extending over half length of bulb (Figs 327–329). Dark, partly sclerotized conductor (C) with round, obtuse, low protuberance on ectal side (Figs 327, 328); short, median apophysis (M) not extending to tip of conductor, in ventral or retrolateral view (Figs 328, 329).

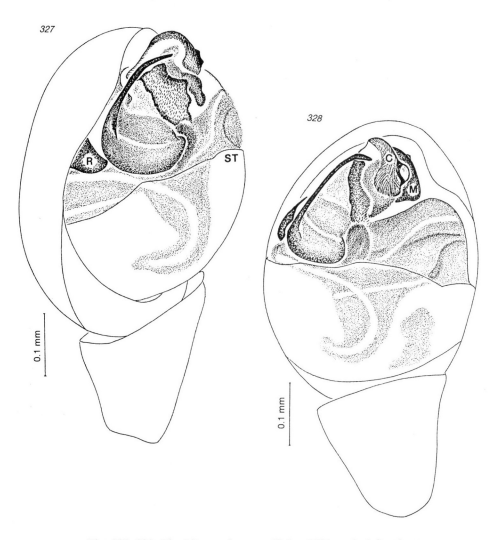

Figs 327–328: *Theridion melanurum* Hahn, 1831; male, left palpus
327. mesal view; 328. ventral view; C — conductor, M — median apophysis,
R — radix, ST — subtegulum

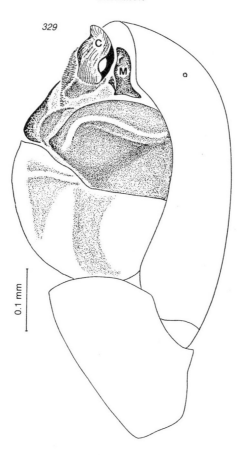

Fig. 329: *Theridion melanurum* Hahn, 1831; male, left palpus,
retrolateral view; C — conductor, M — median apophysis

Female Epigynum: Yellow, sclerotized, central depression extending almost to epigastric furrow (Fig. 330); rims on sides of depression very slightly raised, partly concealing spermathecal orifices. Internal organs form slight swellings outside above central depression (Fig. 330); fine, brown ducts form several loops before entering the large, spermathecal bodies (Fig. 331).

Distirubtion: Holarctis, North Africa, Middle East.

Israel: Throughout the country from Upper Galilee (1), along the Coastal Plain (4–9) and Judean Hills (11) to the sand dunes south of Be'er Sheva' (15).

Specimens are frequently found inside very small webs fastened on the outside of buildings or on tree trunks. Adult males are found from January to April and also in November, and adult females from March to June and again in December.

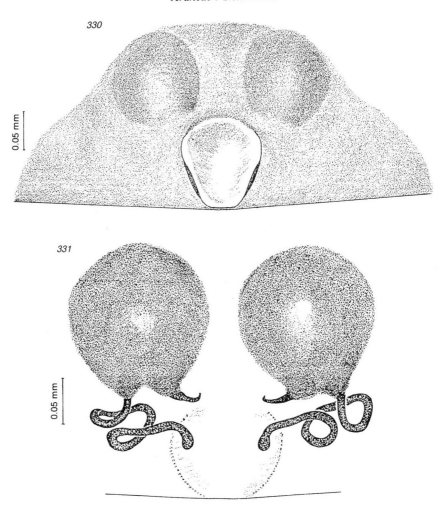

Figs 330–331: *Theridion melanurum* Hahn, 1831; female
330. epigynum; 331. inner spermathecae, dorsal view

Theridion ochreolum Levy & Amitai, 1982
Figs 332–336

Theridion ochreolus Levy & Amitai, 1982a, *J. Zool. Lond.* 196:91; Levy, 1985a, *J. Zool. Lond.* 207:115.

Length of male 1.9–2.8 mm, female 2.1–2.7 mm. Coloration of carapace yellowish-brown with a black mid-dorsal mark and black margins. Sternum in male light brown, in female with blackish, broad dentated band around margins. Legs yellowish with a few black, partly annulated markings on femur and tibia, and dark brown on distal articulations. Opisthosoma in female with light reddish, mid-dorsal, dentated band, light sides with a few black dots and a slender, indistinct white spot on venter. In male, opisthosoma greyish because of white spots mottled with the dark background; posterior part intense white with a few black dots; venter dark brown on anterior swollen portion and with white band below epigastric furrow.

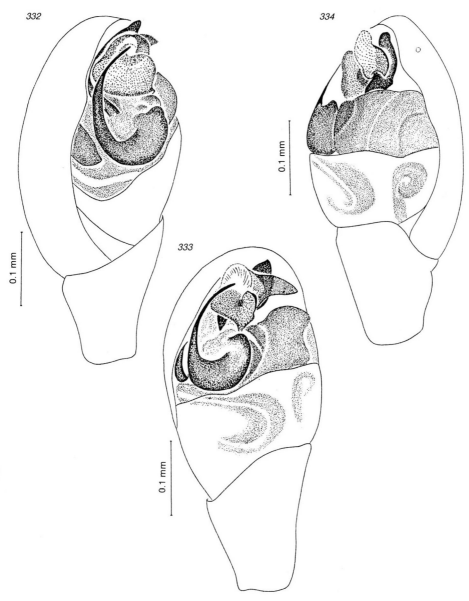

Figs 332–334: *Theridion ochreolum* Levy & Amitai, 1982; male, left palpus
332. mesal view; 333. ventral view; 334. retrolateral view

Male Palpus: Relatively small. Subtegulum extending to about half length of bulb (Figs 332–334). Dark, sclerotic part of conductor extending into a large, conspicuous, horn-like projection, pointing to ectal side (Fig. 333); median apophysis partly hidden by horned extension of conductor, but tip visible above conductor (Figs 332–334).

Female Epigynum: Yellow, transparent, almost rounded central depression separated by a narrow, raised strap from epigastric furrow (Fig. 335); fine, sclerotized rims on posterior sides of depression very slightly raised above orifices of spermathecal tubes. Brown internal spermathecal bodies partly discernible above central depression (Fig. 335); light brown, slightly thickened ducts form one tight loop (Fig. 336).

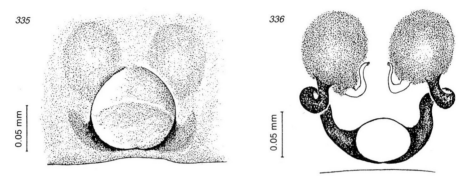

Figs 335–336: *Theridion ochreolum* Levy & Amitai, 1982; female
335. epigynum; 336. inner spermathecae, dorsal view

Distribution: Israel.

Israel: Along the Coastal Plain (4, 9) and Judean Hills (11).

Adults were found from April to June and again in November and December. Spiders are found on walls, both inside and outside of buildings and under the bark of *Eucalyptus* and fruit trees.

Theridion agaricographum Levy & Amitai, 1982
Figs 337–339

Theridion agaricographus Levy & Amitai, 1982a, *J. Zool. Lond.* 196:92.

Male unknown*. Length of female 2,6–2.7 mm. Coloration of carapace brown with dark, broad, median band on dorsum and dark margins. Sternum light with distinct,

---

* *Note added in proof*: A paper by Barbara Knoflach just received contains a description of the unknown male found in Cyprus (*Ber. nat.-med. Verein Innsbruck*, 83: 149–156, Oct. 1996).

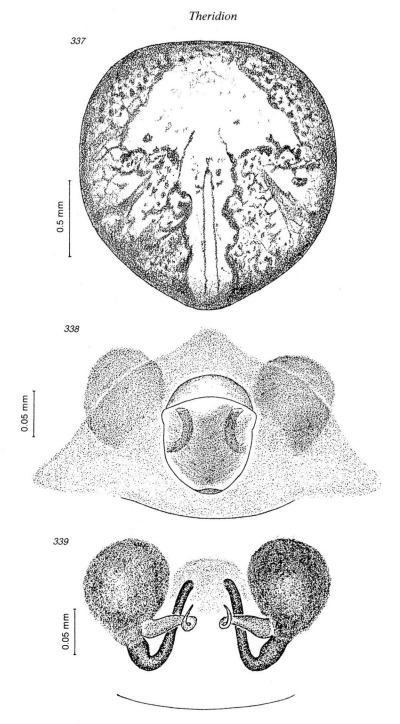

Figs 337–339: *Theridion agaricographum* Levy & Amitai, 1982; female
337. opisthosoma, dorsal surface; 338. epigynum; 339. inner spermathecae, dorsal
view

179

dark margins. Legs light brown with a few dark spots. Opisthosoma grey with a conspicuous, white, mushroom-like pattern on back: wide in front and a slender, scalloped stalk extending backwards (Fig. 337); venter white except for a black marking in front of spinnerets and black patch on both sides of pedicel.

Female Epigynum: Yellowish central depression partly covered anteriorly by fine, transparent hood (Fig. 338); internal spermathecae partly visible through integument. Ducts of spermathecae recurve on themselves and extend to about two-thirds the height of spermathecal bodies (Fig. 339).

Distribution: Israel.

Israel: 'Avdon in the western Galilee (1) and Jerusalem (11).

Adult females were found in April.

Theridion vespertinum Levy, 1985
Figs 340–342

Theridion vespertinus Levy, 1985a, *J. Zool. Lond.* 207:119.

Length of male 2.6 mm, female unknown. Carapace brown, transparent with a black median indentation (=fovea). Sternum uniformly black. Legs brown, mottled with black spots. Opisthosoma greyish-black with a light, mid-dorsal, dentated band; edges of band marked with black; sides of opisthosoma covered by a black patch merging in part with the grey background; venter black, swollen anteriorly, centre light, and spinnerets encircled by a black ring.

Male Palpus: Subtegulum (ST) extending to about half length of bulb. Large, broad black radix (R), at middle of mesal side, obliquely protruding towards tip of bulb (Fig. 340). Short filiform portion of embolus (E) rising from proximal side of basal embolar division (Fig. 341). Dark, partly sclerotized conductor (C) with a low, pointed protrusion on ectal side (Figs 341, 342). Stout, large median apophysis (M) readily apparent inside back of bulb (Figs 341, 342).

Distribution: Israel, known only from the type locality, Jerusalem (11).

The one male known thus far was found at night on a wall inside a building in January.

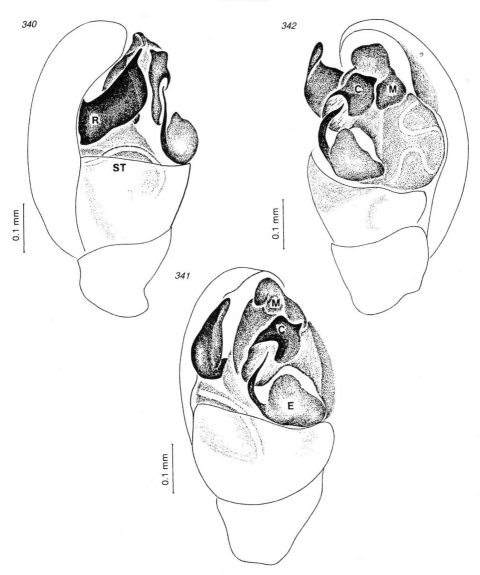

Figs 340–342: *Theridion vespertinum* Levy, 1985; male, left palpus
340. mesal view; 341. ventral view; 342. retrolateral view; C – conductor, E – embolus,
M – median apophysis, R – radix, ST – subtegulum

Theridion simile C.L. Koch, 1836
Figs 343–348

Theridion simile C.L. Koch, 1836, *Die Arachniden*, 3:62, pl. 215, Nürnberg; Roewer, 1942,
Katalog der Araneae 1:471; Levi, 1957, *Bull. Am. Mus. nat. Hist.* 112:53; Bonnet, 1959,
Bibliographia Araneorum 2(5):4525; Levy & Amitai, 1982a, *J. Zool. Lond.* 196:94.

Length of male 1.8–3.0 mm, female 2.2–3.2 mm. Coloration of carapace light yellow, almost transparent, with red-brown, broad, mid-dorsal band. Sternum of female, yellowish; in male, with brown margins. Legs whitish with brown articulations. Opisthosoma on centre of back of female, usually, with large yellow rhomboidal blotch, encircled by dark, brown mottled with white dots and reddish stripes (Fig. 343); posterior part of dorsum with yellowish-green median band, tapering towards

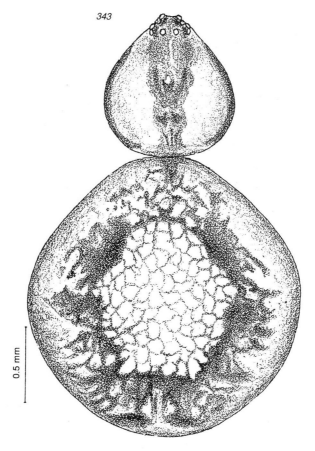

Fig. 343: *Theridion simile* C.L. Koch, 1836; female,
dorsal view of spider

spinnerets; sides of opisthosoma greenish with two brown bars; venter greenish with few white dots traversed by brown band in front of spinnerets. Opisthosoma of male, red-brown with indistinct, light central spot on dorsum; sides greenish; venter with brown, broad longitudinal band extending from pedicel to epigastric furrow, and brown bar traversing below furrow.

Male Palpus: Very small, with distal parts sometimes translucent and hardly visible. Filiform portion of embolus concealed between membranous folds of conductor (C) and median apophysis (M, Figs 344, 345); fine lancet-like extension of conductor (L) tightly appressed to apical structure of bulb, visible on ectal side (Fig. 346).

Female Epigynum: Central depression on plate slightly removed from epigastric furrow; posterior borders of depression slightly raised and funnel-like (Fig. 347). Internal spermathecae visible through integument (Fig. 347). Short ducts of small spermathecae extending only to sides of central depression (Fig. 348).

Distribution: Holarctic, North Africa, Israel and probably Lebanon.

Israel: From Upper Galilee (1) to the Judean Hills (11).

Adults of both sexes are found mainly in April and May. Specimens are found inside small irregular webs spun at tips of twigs of various plants. The webs are made out of very fine threads. The spider usually retreats to the underside of a leaf on the slightest disturbance.

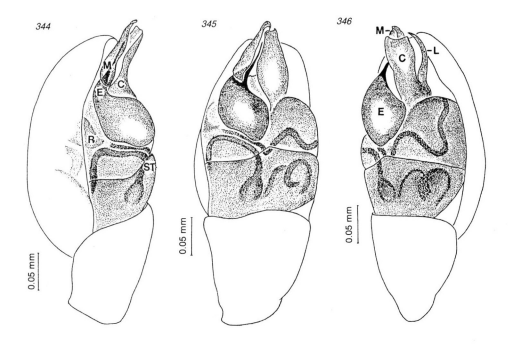

Figs 344–346: *Theridion simile* C.L. Koch, 1836; male, left palpus
344. mesal view; 345. ventral view; 346. retrolateral view; C – conductor, E – embolus,
L – lancet-like extension, M – median apophysis, R – radix, St – subtegulum

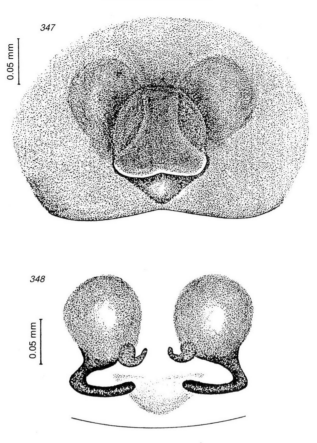

Figs 347–348: *Theridion simile* C.L. Koch, 1836; female
347. epigynum; 348. inner spermathecae, dorsal view

Theridion accoense Levy, 1985
Figs 349–352

Theridion accoensis Levy, 1985a, *J. Zool. Lond.* 207:117.

Length of male 1.7 mm, female unknown. Carapace light yellow with dark posterior sides and a black, mid-dorsal mark slightly attenuated in front (Fig. 349). Sternum yellow with margins encircled by a fine black line. Legs yellowish with black, partly annulated marking on femora and close to distal articulations. Opisthosoma white dorsally, along middle, and yellow upper sides traversed by rows of black, conspicuous dots (Fig. 349); areas surrounding pedicel black, and posterior end of opisthosoma marked by a few black spots laterally extended; venter light at centre and spinnerets not encircled with black.

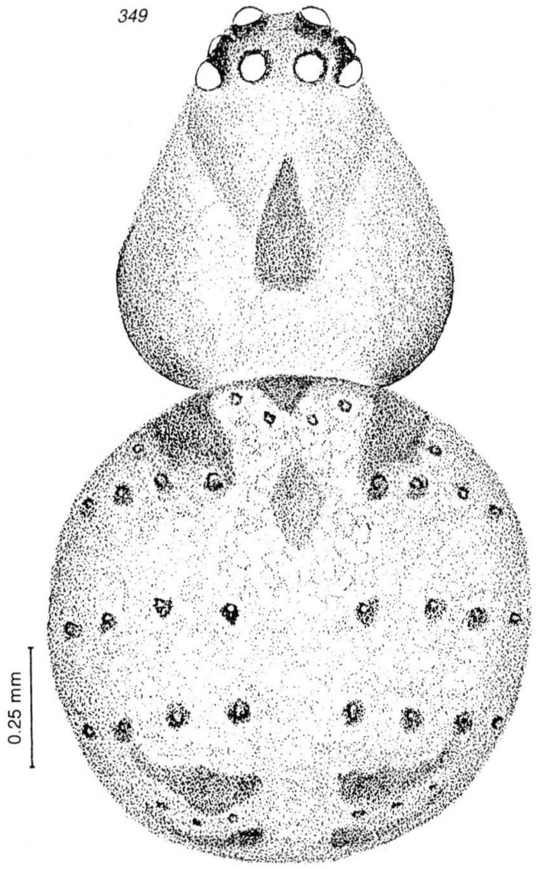

Fig. 349: *Theridion accoense* Levy, 1985; male,
dorsal view of spider

Male Palpus: Very small. Black, thick embolus (E) placed on mesal part of bulb (Figs 350, 351); embolus tapers distally but has no filiform portion. Short, stubbed radix (R) visible on middle of mesal side. Membranous conductor (C) rises at centre (Figs 351, 352); contours of its distal, translucent part are barely traceable and it blurs the view of the median apophysis (M) rising at back (Fig. 351); latter best viewed from retrolateral side (Fig. 352).

Distribution: Israel, known only from the type locality, 'Akko (=Acco, Acre), western Galilee (4).

The one male known as yet was found at the end of November, in an avocado plantation.

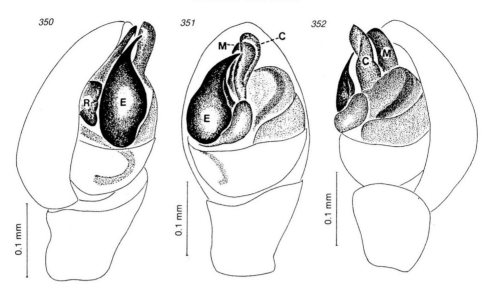

Figs 350–352: *Theridion accoense* Levy, 1985; male, left palpus
350. mesal view; 351. ventral view; 352. retrolateral view; C – conductor, E – embolus,
M – median apophysis, R – radix

Theridion hemerobius Simon, 1914
Figs. 353, 354

Theridion hemerobius Simon, 1914: 264; Roewer, 1942, *Katalog der Araneae* 1:464; Bonnet,
1959, *Bibliographia Araneorum* 2(5):4478; Levy & Amitai, 1982a, *J. Zool. Lond.* 196:96
(provisionally as *T. ? pictum*); Bosmans, Vanuytven & Van Keer, 1994, *Bull. Br. arachnol.
Soc.* 9(7):236, figs 1–5.

Male as yet unknown from Israel. Length of female 3.10–3.25 mm. Coloration of
carapace yellowish with dark, broad median band, slightly constricted in middle,
extending along entire length; margins with a distinct dark line. Sternum light. Legs
light brown with dark markings mainly near articulations. Opisthosoma grey-brown
with light, distinct wide dentated median band on back; venter light.
Female Epigynum: Posterior margin of epigynal plate along epigastric furrow, raised
and folded (Fig. 353). Central opening very small, slightly oval, with fine, indistinct
sclerotized rims (Fig. 353). Ducts of large, brown spermathecae recurve slightly (Fig.
354).
Distribution: Western Europe to Bulgaria, Corsica, Israel.
Israel: Lake Hula, northern Galilee (1).
Only two adult females were collected in April, 1940 on heads of papyrus reed
(*Cyperus*). According to Bosmans *et al.* (1994) these females should be placed in

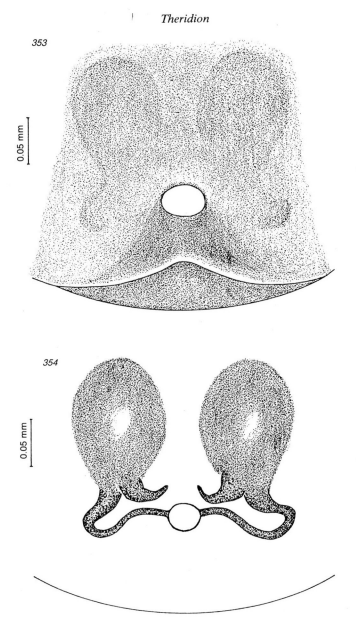

Figs 353–354: *Theridion hemerobius* Simon, 1914; female
353. epigynum; 354. inner spermathecae, dorsal view

T.hemerobius, however, this identification must remain provisional unless substantiated by males collected in Israel.

Theridion hermonense Levy, 1991
Figs 355, 356

Theridion hermonense Levy, 1991, *Bull. Br. arachnol. Soc.* 8(7):227.

Male unknown. Length of female 1.1–1.3 mm. Coloration generally light red with no pattern on back or venter of opisthosoma.

Female Epigynum: Concave epigynal plate with brown, slightly raised fold posteriorly along borders of epigastric furrow (Fig. 355); touching, black encircled, small outer orifices of winding inner ducts placed at middle of epigynal plate far from epigastric furrow. Brown internal spermathecal bodies discernible through transparent integument (Fig. 355); coils of winding ducts lie dorsal to spermathecal bodies, not surrounding them (Fig. 356).

Distribution: Israel: Mt Hermon, 1950 m (19).

Two adult females, considered the smallest *Theridion* of Israel, were found running on the underside of a stone. They were collected in July at the bottom of a large doline with spiny cushion shrubs (tragacanthic vegetation).

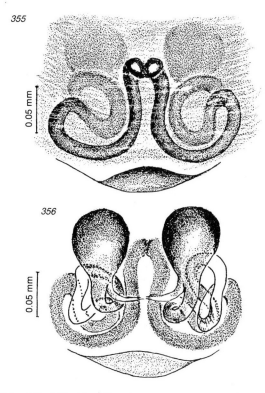

Figs 355–356: *Theridion hermonense* Levy, 1991; female
355. epigynum; 356. inner spermathecae, dorsal view

Theridion nigropunctatum Lucas, 1846

Figs 357–363

Theridion nigropunctatum Lucas, 1846, *Explor. scient. Algér. Zool.* 1:266, pl. 16, fig. 6;
  Roewer, 1942, *Katalog der Araneae*, 1:467; Bonnet, 1959, *Bibliographia Araneorum*,
  2(5):4494; Levy & Amitai, 1982a, *J. Zool. Lond.* 196:97.

Length of male 2.9 mm, female 3.1–3.4 mm. Carapace yellowish with black mid-dorsal
band, black markings on sides, dark line around margins and a small black spot, in
front, centrally on edges of clypeus. Sternum light with conspicuous, black spot in the
middle and dark dentated band around margins (Fig. 357). Legs light, mottled with
black dots and a few black markings near articulations. Opisthosoma mottled
throughout with many small, some slightly larger, black dots on light background
(Fig. 358); dorsum centrally traversed by white band; venter white with one median
black spot and six black, short bars, around spinnerets.

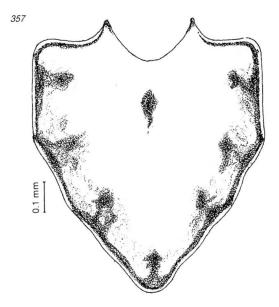

357

0.1 mm

Fig. 357: *Theridion nigropunctatum* Lucas, 1846; female,
sternum, ventral surface

Male Palpus: Radix hardly proturding on mesal side (Fig. 359). Short, filiform portion
of embolus extends distally to fleshy, long and slender conductor (Figs 359–361).
Brown, large median apophysis visible on back, inside (Figs 359–361).
Female Epigynum: Central depression on epigynal plate bordered anteriorly by dark,
broad band coiling strongly inwards on sides (Fig. 362); posterior portion of central
depression, along epigastric furrow, only slightly raised with indistinct rims. Small

189

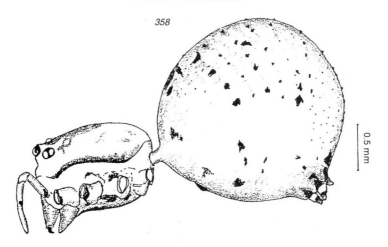

Fig. 358: *Theridion nigropunctatum* Lucas, 1846; female,
lateral view of spider (legs omitted)

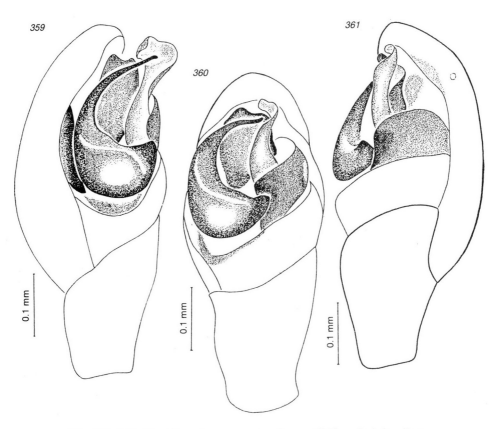

Figs 359–361: *Theridion nigropunctatum* Lucas, 1846; male, left palpus
359. mesal view; 360. ventral view; 361. retrolateral view

190

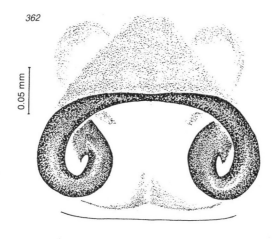

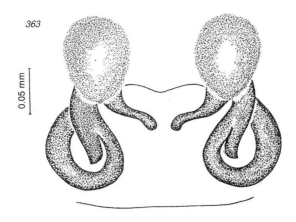

Figs 362–363: *Theridion nigropunctatum* Lucas, 1846; female
362. epigynum; 363. inner spermathecae, dorsal view

yellow brown spermathecae with thick, brown peculiarly coiled tubes from which a short stub branches off (Fig. 363).

Distribution: Mediterranean countries.

Israel: Along the Coastal Plain (4, 8) to the Judean Hills (11).

Adult males were collected in January and adult females were found in March and May. Specimens are rather conspicuously marked; apparently not very common. Some were taken on pine trees (*Pinus halepensis*).

Theridion melanostictum O.P.-Cambridge, 1876
Figs 364–369

Theridion melanostictum O.P.-Cambridge, 1876, *Proc. zool. Soc. Lond.* 1876:570; Roewer, 1942, *Katalog der Araneae*, 1:466; Bonnet, 1959, *Bibliographia Araneorum*, 2(5):4490; Levy & Amitai, 1982a, *J. Zool. Lond.* 196:99; Levy, 1985a, *J. Zool.* 207:114.

Length of male 2.3–2.5 mm, female 2.2–3.6 mm. Coloration of carpace light brown with, sometimes indistinct, dark, mid-dorsal band and dark margins. Sternum evenly dark. Legs yellow-brown with dark markings near articulations. Opisthosoma yellow mottled with white, black and reddish spots; some specimens have on dorsum a white, median, dentated band; venter traversed in middle by large, greenish-brown to black patch continuing as a dark belt on the sides dorsally; sometimes venter light with one black spot in front of spinnerets and two black, oblique markings, behind spinnerets; venter, in male, also black and markedly swollen anteriorly.

Male Palpus: Relatively large. Radix, on middle of mesal side, finger-like, distinctly protruding (Fig. 364). Filiform portion of embolus rising distally from rounded,

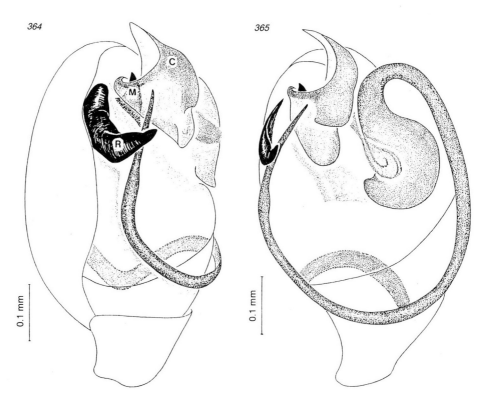

Figs 364–365: *Theridion melanostictum* O.P.-Cambridge, 1876; male, left palpus
364. mesal view; 365. ventral view; C — conductor, M — median apophysis, R — radix

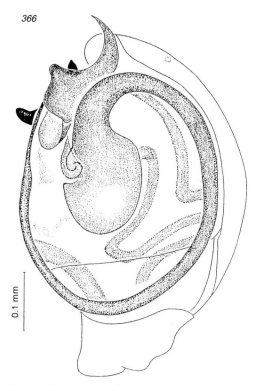

366

0.1 mm

Fig. 366: *Theridion melanostictum* O.P.-Cambridge, 1876; male, left palpus,
retrolateral view

embolar basal division (Figs 365, 366); embolar duct encircling most of bulb, almost
touching tibia proximally (Figs 365, 366); pointed median apophysis almost entirely
hidden by circular extension of conductor (Figs 364, 365); tip of elaborate, sclerotic
conductor tapering above cymbium (Figs 364–366).

Female Epigynum: Central depression of epigynal plate often indistinct, sometimes
margins on sides of depression slightly raised by bulging internal tubes (Figs 367, 368);
orifices of ducts open close to epigastric furrow; openings may be separated (Fig. 367)
or, touching centrally and appearing as one, short reddish bar (Fig. 368). Internal
organs excccdingly claboratc: slcndcr, clongated spermathecae surrounded by long,
looping ducts, forming double coiled helices (Fig. 369).

Distribution: USA (Florida), Egypt, Israel.

Israel: Coastal Plain (4, 8) and along the Rift Valley (7, 13).

Adults are found on plants possibly throughout the year. A female with a spherical,
tough coated egg sac, 2.5 mm in diameter, was collected near the Dead Sea in October.
The sac contained about 40 eggs, 0.5 mm in diameter. The population found along the
Mediterranean coast displays larger slightly more colourful females (3.0–3.6 mm in
length) than specimens of the Rift Valley population (females attain 2.2–3.0 mm in

length). The disjunct distribution: Middle East and America, may result from human introduction or reflect gaps in our knowledge; sometimes the missing link in distribution is hidden under a misidentification (Levy, 1985a:114).

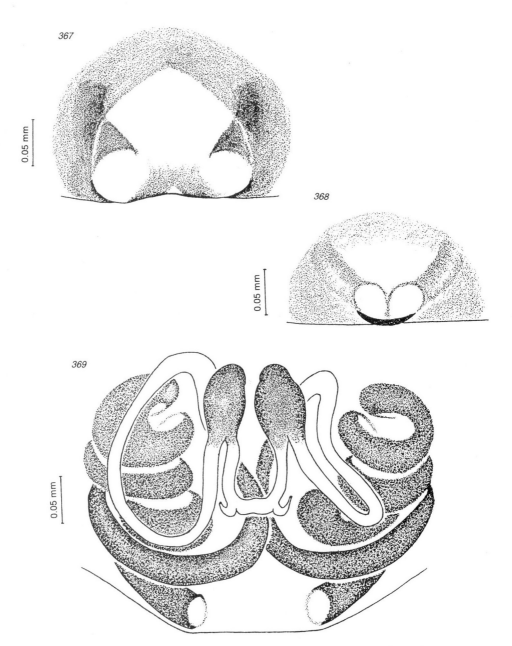

Figs 367–369: *Theridion melanostictum* O.P.-Cambridge, 1876; female
367. epigynum; 368. epigynum, variation; 369. inner spermathecae, dorsal view

Theridion hierichonticum Levy & Amitai, 1982
Figs 370–372

Theridion hierichonticus Levy & Amitai, 1982a, *J. Zool. Lond.* 196:102.

Length of male 2.1 mm, female unknown. Carapace light brown with slightly darker markings on centre and sides. Sternum light with dark margins. Legs brown. Opisthosoma on dorsum with faintly indicated white, median, dentated band and a series of black spots arranged along each side of band; sides and central space of venter, creamy, mottled with white spots; anterior portion of venter dark and swollen, posteriorly with two distinct, black spots on each side of spinnerets.

Male Palpus: Radix relatively small, lobe-like, projecting at about middle of mesal side of bulb (Fig. 370). Filiform portion of embolus rising from proximal side of basal embolar division, curving close to edges of subtegulum (Figs 371, 372); stout, cone-shaped median apophysis readily apparent on mesal side of bulb (Fig. 370). Beaked tip of sclerotic conductor extending to about height of cymbium.

Distribution: Israel, known only by the male holotype collected in March near Jericho (13).

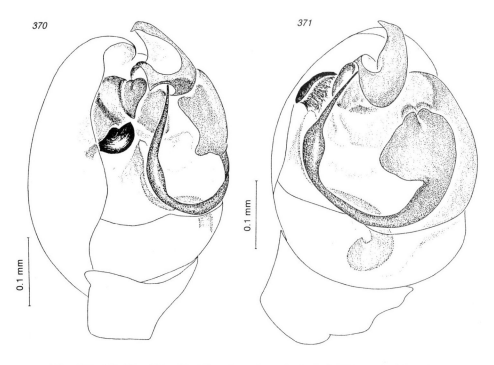

Figs 370–371: *Theridion hierichonticum* Levy & Amitai, 1982; male, left palpus
370. mesal view; 371. ventral view

195

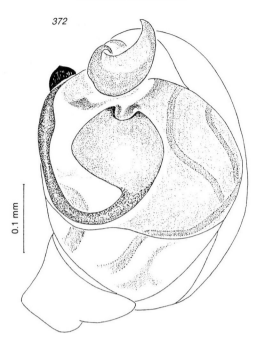

372

0.1 mm

Fig. 372: *Theridion hierichonticum* Levy & Amitai, 1982; male, left palpus,
retrolateral view

Theridion jordanense Levy & Amitai, 1982
Figs 373, 374

Theridion jordanensis Levy & Amitai, 1982a, *J. Zool. Lond.* 196:103.

Male unknown. Length of female 1.9–2.6 mm. Carapace yellowish with black, mid-
dorsal band and brown eye tubercles. Sternum light with black line around margins.
Legs light with a few fine dark markings. Opisthosoma grey; dorsum with white
patches tinged with red along median, faint yellow line; on back, with four distinct
black spots; three large, conspicuous, additional black spots around spinnerets: one on
venter in front and two obliquely behind spinnerets; venter reddish-brown, traversed
centrally by fine, light line.
Female Epigynum: Epigynal plate cone-like markedly raised, very distinct in profile.
Only two round orifices with slightly sclerotized walls, visible on top of cone (Fig. 373);
placement of internal spermathecal bodies readily visible through integument on
anterior slope of raised epigynum. Dark, brown, pear-shaped spermathecae with
strongly convoluting ducts converging centrally (Fig. 374).
Distribution: Israel, possibly Jordan.
Israel: Around Lake Kinneret (7) and near the Dead Sea (13).

196

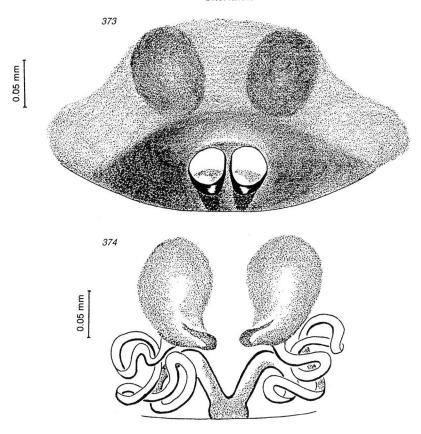

Figs 373–374: *Theridion jordanense* Levy & Amitai, 1982; female
373. epigynum; 374. inner spermathecae, dorsal view

Adult females were found on plants in June–July and in September; in the latter month, together with young. It is possible that *T. jordanense* is actually the male of *T. hierichonticum* but there is no evidence for it.

Theridion musivum Simon, 1873
Figs 375–379

Theridium musivum Simon, 1873, *Mém. Soc. r. Sci.* Liège (2)5:94, pl. 2, fig. 26; Roewer, 1942, Katalog der Araneae, 1:467; Bonnet, 1959, *Bibliographia Araneorum* 2(5):4493; Levy & Amitai, 1982a, *J. Zool. Lond.* 196:105.

Male as yet unknown from the Middle East. Length of female from Sinai 1.7 mm. Clypeal region of carapace slightly concave. Sternum convex, almost rounded, wider than long. Coloration of opisthosoma grey with some dusky markings on back.

197

Male Palpus: Drawings provided (Figs 375–377) are of a male from Italy.

Female Epigynum: Central depression on epigynal plate, bordering epigastric furrow, with slightly sclerotized, raised rims (Fig. 378); adjoining orifices of dark, thick-walled, diverging tubes visible inside central depression (Fig. 378). Internal spermathecal

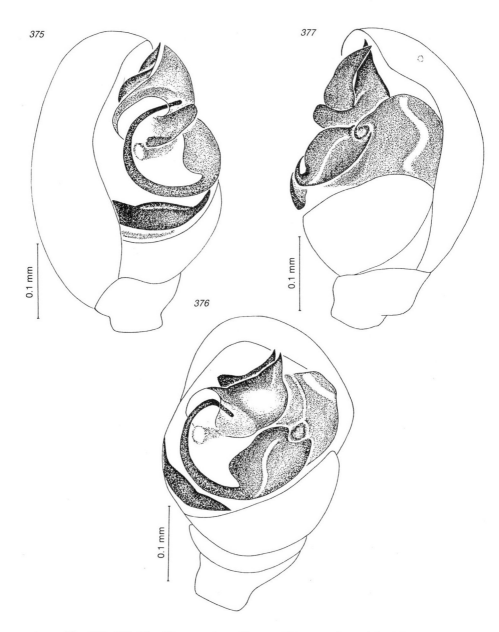

Figs 375–377: *Theridion musivum* Simon, 1873; male from Italy, left palpus
375. mesal view; 376. ventral view; 377. retrolateral view

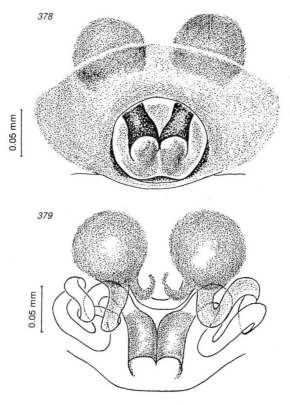

Figs 378–379: *Theridion musivum* Simon, 1873; female from Sinai
378. epigynum; 379. inner spermathecae, dorsal view

organs partly discernible through transparent integument. Coiled ducts of reddish-brown, rounded spermathecae, with black, very thick and widened portion close to external entrances of ducts (Fig. 379).

Distribution: Spain, France, Italy, Morocco, Algeria, Egypt (Sinai), possibly Israel.

Egypt (Sinai): Qadesh Barnea' (21), November.

Theridion negebense Levy & Amitai, 1982
Figs 380–387

Theridion negebensis Levy & Amitai, 1982a, *J. Zool. Lond.* 196:106.

Length of male 2.1–2.6 mm, female 2.3–3.5 mm. Coloration of carapace light brown to brown with black median indentation (=fovea) and dark sides. Sternum uniformly dark. Legs yellowish-brown mottled with dark markings. Opisthosoma with white, conspicuous, median dentated band along entire back; margins of band marked with

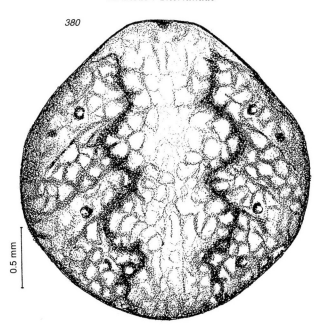

380

0.5 mm

Fig. 380: *Theridion negebense* Levy & Amitai, 1982; female,
opisthosoma, dorsal surface

intense deep brown (Fig. 380); sides of opisthosoma creamy, with brown band arching
from in front downwards to middle of venter, and a few white and brown dots on
posterior part; venter with dark, broad band, slightly constricted by two white blotches,
below epigastric furrow.

Male Palpus: Black, large radix extending over most of mesal side of bulb, with
proximal portion strongly bent and ending with pointed tip distally directed (Figs 381,
382). Filiform portion of embolus rising from retrolateral side of basal division of
embolus (Figs 382, 383); embolar duct partly concealed on mesal side by transparent,
elongate, membranous fold (Figs 381, 382); elaborate, sclerotic structure of conductor
not extending above cymbium (Figs 381–383).

Female Epigynum: Raised, swollen epigynal plate most distinct in profile. Large,
slightly oval, central depression, surrounded by dark, sclerotic rims, widening cup-like
on lower corners of posterior margin (Fig. 384); orifices of ducts partly hidden. Course
of winding ducts of spermathecae varies slightly in different specimens but coils always
remain in space between spermathecae and epigastric furrow, not surrounding bodies
of spermathecae (Figs 385–387); left and right sides usually not symmetrical.

Distribution: Southern Israel.

Israel: North and Central Negev (15, 17).

Adults are found in May and in August–September. An egg sac laid in May was
3.4 mm in diameter and contained a few dozen eggs about 0.5 mm in diameter.

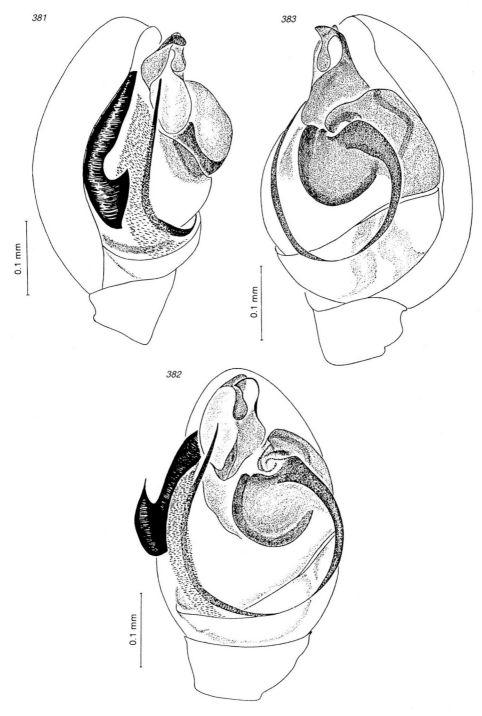

Figs 381–383: *Theridion negebense* Levy & Amitai, 1982; male, left palpus
381. mesal view; 382. ventral view; 383. retrolateral view

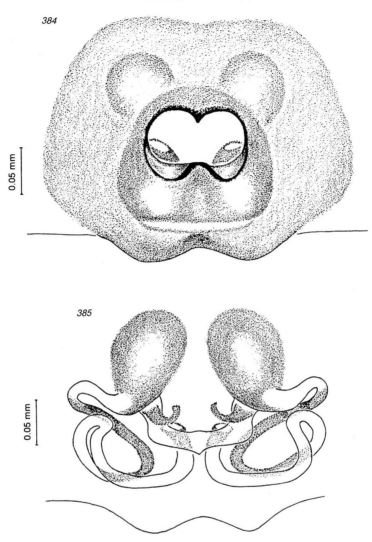

Figs 384–385: *Theridion negebense* Levy & Amitai, 1982; female
384. epigynum; 385. inner spermathecae, dorsal view

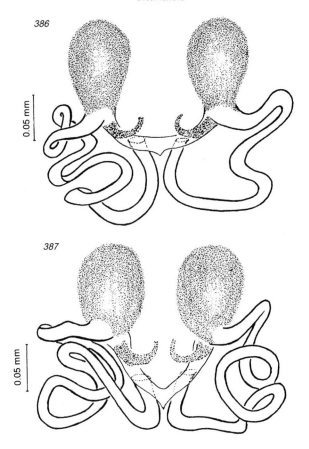

Figs 386–387: *Theridion negebense* Levy & Amitai, 1982; female
386. inner spermathecae, dorsal view, variation; 387. inner spermathecae, dorsal view, variation

Theridion gekkonicum Levy & Amitai, 1982
Figs 388, 389

Theridion gekkonicus Levy & Amitai, 1982a, *J. Zool. Lond.* 196:109.

Male unknown. Length of female 2.1–2.4 mm. Coloration of carapace blackish, sternum evenly black. Legs light with distinct, black annulations. Opisthosoma dorsally with white and dark spots on deep brown background and a median dentated band extending to spinnerets; sides dusky with large white spot in centre; venter black with two small, conspicuous white spots below epigastric furrow and four additional white spots: two on each side of spinnerets.

Female Epigynum: Central depression in epigynal plate almost entirely encircled by dark, sclerotized rims (Fig. 388); bottom of central depression with broad, median

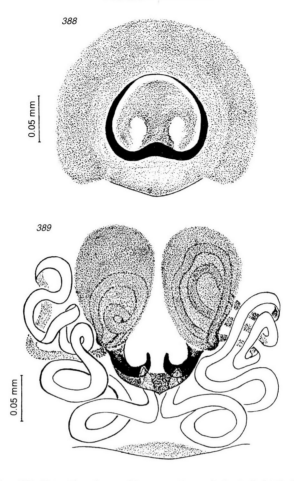

Figs 388–389: *Theridion gekkonicum* Levy & Amitai, 1982; female
388. epigynum; 389. inner spermathecae, dorsal view

septum-like swelling; slits of spermathecal orifices visible on sides of septum (Fig. 388). Internal organs not discernible. Dark, large, elongated spermathecae partly enveloped by long, strongly coiled ducts (Fig. 389).
Distribution: Israel.
Israel: Near Lake Kinneret (7) and in Coastal Plain (8).
Adult females are found in March–April.

Theridion dromedarius Simon, 1880

Figs 390–396

Theridion dromedarius Simon, 1880, *Annls soc. entom. Fr.* (5) 10:99; Bonnet, 1959, *Bibliograhia Araneorum* 2(5):4469; Levy & Amitai, 1982a, *J. Zool. Lond.* 196:110.

Length of male 1.7–1.9 mm, female 1.65–1.95 mm. Coloration in general yellow to orange brown. Carapace in male with black median indentation (Fig. 390), in female with large brown band only on anterior dorsal part and black line surrounding margins. Male with stridulatory ridges on posterior part of carapace. Light brown sternum slightly obtuse anteriorly. Legs light brown with dark annulations. Opisthosoma on dorsum, in front, with two sideways projecting, coned tubercles, less distinct in male (Fig. 391); region above pedicel in male, with fine denticles. Opisthosoma dorsally, orange-brown slightly mottled with black and white in front of tubercles,

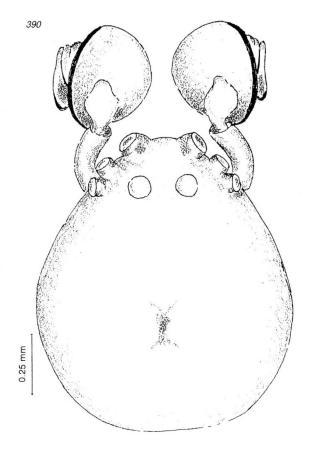

390

0.25 mm

Fig. 390: *Theridion dromedarius* Simon, 1880; male,
carapace and palpi, dorsal view

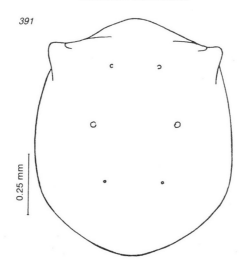

Fig. 391: *Theridion dromedarius* Simon, 1880; female,
opisthosoma, dorsal view

yellow-orange with white along a transversal line connecting both tubercles, and with orange-brown pattern tapering posteriorly to above spinnerets; venter reddish-brown. *Male Palpus*: Bulb is rotated outwards with back of cymbium turned mesally (Fig. 390). Embolar duct encircling one and a half times surface of ectal side of bulb (Figs 392–394); tip of embolar duct extending distally to leaf-like, elongated, membranous conductor (Figs 392–394).

Female Epigynum: Epigynal plate with a wide, central opening extending along epigastric furrow (Fig. 395). Dusky and whitish portions of internal organs discernible through integument (Fig. 395). Dark spermathecae, obliquely placed, with deep constriction in middle (Fig. 396); ducts coiled tightly around constriction and widen gradually towards external opening (Fig. 396).

Distribution: Canary Islands, Spain, throughout northern Africa, Yemen, Israel.

Israel: In the Central Coastal Plain (8) as well as near Lake Kinneret (7), the southern Negev and Elat (16).

Adults were collected in March, June, September and October. They are found on trees and on several occasions were taken from small, 10 cm in diameter, perfect orbital webs suspended among lower branches. Mistaken for Araneidae, the true identity of these tiny spiders was discovered only under high power magnification in the laboratory; their association, if any, with the orbital webs has thus not yet been elucidated.

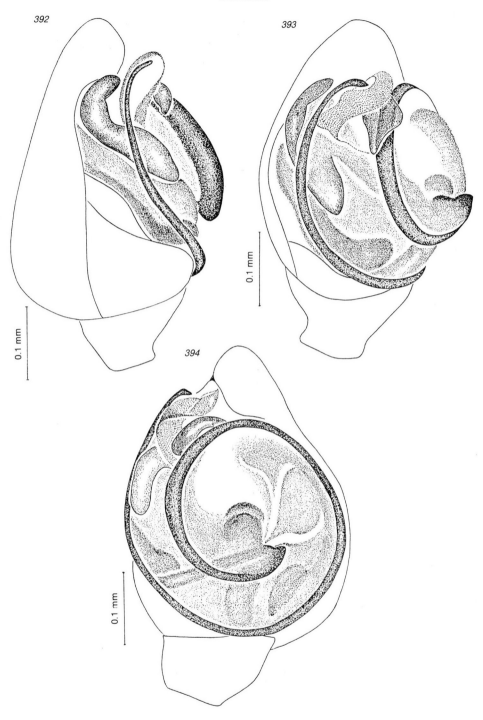

Figs 392–394: *Theridion dromedarius* Simon, 1880; male, left palpus
392. mesal view; 393. ventral view; 394. retrolateral view

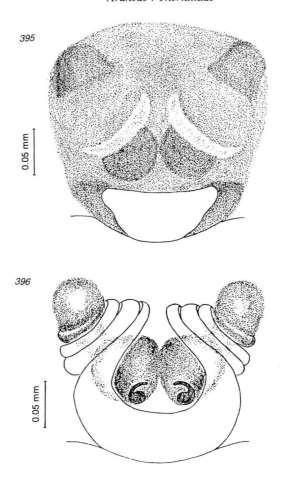

Figs 395–396: *Theridion dromedarius* Simon, 1880; female
395. epigynum; 396. inner spermathecae, dorsal view

Theridion dafnense Levy & Amitai, 1982
Figs. 397, 398

Theridion dafnensis Levy & Amitai, 1982a, *J. Zool. Lond.* 196:113.

Male unknown. Length of female 2.7 mm. Coloration of carapace black. Sternum uniformly dark. Legs light with dark annulations. Opisthosoma black in front, above pedicel and with conspicuous mid-dorsal, dentated band, most distinct on posterior portion; sides and venter light, gradually darkening posteriorly.

Female Epigynum: Yellowish epigynal plate with central depression bordering epigastric furrow (Fig. 397); large, brown, semicircular, pit-like openings, partly separated by inwardly, tapering projection, placed under anterior borders of central depression

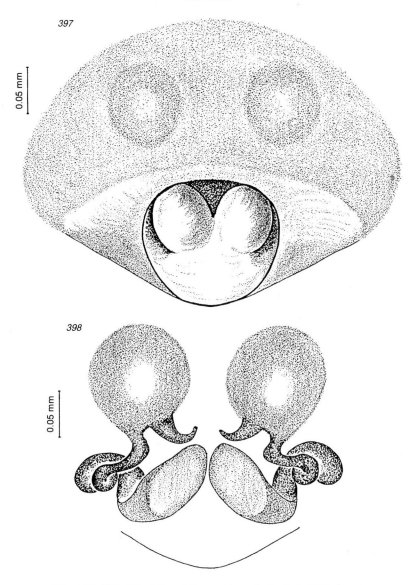

Figs 397–398: *Theridion dafnense* Levy & Amitai, 1982; female
397. epigynum; 398. inner spermathecae, dorsal view

(Fig. 397); entrances to spermathecal ducts actually hidden on posterior sidewalls of external pits. Round, brown spermathecae with partly slender, coiled ducts and a much widened yellow, finely sclerotized, transparent portion close to external entrances (Fig. 398).

Distribution: Israel, known so far only from the type locality, Dafna, in the northeastern Galilee (1), May.

Theridion vallisalinarum Levy & Amitai, 1982
Figs 399, 400

Theridion vallisalinarum Levy & Amitai, 1982a, *J. Zool. Lond.* 196:115.

Male unknown. Length of female 2.1 mm. Coloration of carapace and sternum light, both with a black line around margins. Legs light with dark markings close to articulations. Opisthosoma on dorsum with distinct, white median dentated band extending to spinnerets; edges of band marked with black; sides creamy, mottled with white spots; venter light with dark chevron-like mark, with apex pointing to spinnerets. *Female Epigynum*: Central depression on epigynal plate partly surrounded by thick, funnel-like, semicircular rims (Fig. 399); depression plugged with stopper-like, hard reddish and opaque matter. Spermathecae small, slender and slightly elongated (Fig.

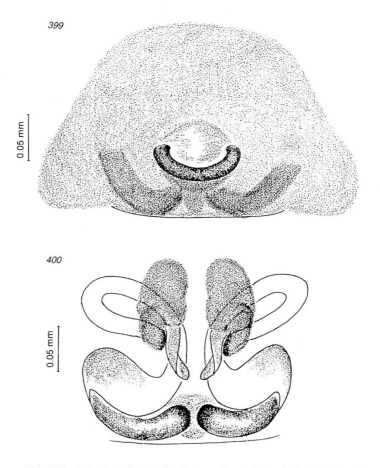

Figs 399–400: *Theridion vallisalinarum* Levy & Amitai, 1982; female
399. epigynum; 400. inner spermathecae, dorsal view

400); ducts forming one large loop beyond spermathecae and one strongly recurved portion in form of thick, widened bend, close to epigastric furrow (Fig. 400).

Distribution: Israel, known only from the type locality, Kallia, on the northern shores of the Dead Sea (13), April.

Theridion uncinatum Lucas, 1846
Figs 401–407

Theridion uncinatum Lucas, 1846, *Explor. scient. Algér. Zool.* 1:267, pl. 17, fig.2; Levy & Amitai, 1982a, *J. Zool. Lond.* 196:116.

Length of male 3 mm, female 3.2–3.6 mm. Coloration of the two sexes differs entirely. In male, carapace and sternum deep brown. Bulbs of palpi black, slender femora brown. Legs intense yellow except femora; femora of first legs with black stripe along

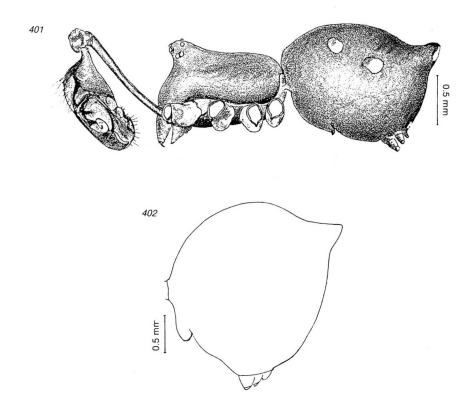

Figs 401–402: *Theridion uncinatum* Lucas, 1846
401. male, lateral view of spider (legs omitted);
402. female, opisthosoma, lateral view

211

frontal side and femora of third and fourth pairs, black on posterior side. Opisthosoma of male black with six conspicuous white spots and a strong, dorso-posterior tubercle (Fig. 401); four spots placed on upper parts of sides of opisthosoma, remaining two spots, placed one under posterior projecting tubercle, the other, close behind spinnerets. In female, carapace yellow brown with dark marking on centre and sides. Clypeus projecting obliquely forward. Sternum light brown. Legs light yellow. Opisthosoma with posterior, backward directed, dark brown prominent conical tubercle (Fig. 402); dorsum whitish-yellow with two parallel rows of four pairs of distinct, intense white spots; sides with a few bright white stripes and white spots placed among them; one stripe also on posterior part, under conical tubercle; venter with one white blotch below epigastric furrow, and four small white spots surrounding spinnerets.

Male Palpus: Very large, kidney-bean shaped, held vertically in front of spider. Size of bulb including the small tibia larger than entire frontal height of prosoma together with the chelicerae (Fig. 401). Large, leaf-like rounded, light coloured radix protruding on mesal side and partly enveloping dark, thick, finger-like median apophysis (Figs 403, 404); deep brown embolus forming large, bulging curve in middle of bulb, and

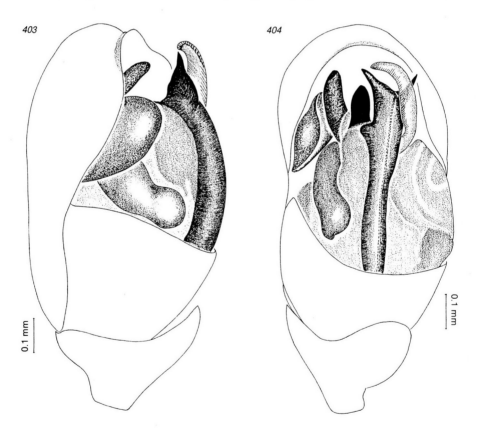

Figs 403–404: *Theridion uncinatum* Lucas, 1846; male, left palpus
403. mesal view; 404. ventral view

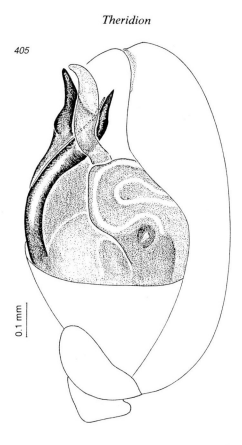

405

0.1 mm

Fig. 405: *Theridion uncinatum* Lucas, 1846; male, left palpus,
retrolateral view

apical part, splitting into sharp pointed, sclerotic tips (Figs 403–405); stout, black accessorial protrusion of embolus, projecting apically, between median apophysis and pointed tips of embolus (Fig. 404); light, opaque, fleshy and slender conductor visible on retrolateral side (Fig. 405).

Female Epigynum: Deep brown, relatively very large epigynal plate usually covered by hardened secretion. Central depression with fine, raised rims, converging close to epigastric furrow (Fig. 406); black slits of spermathecal ducts visible anteriorly, on sides of epigynum, and thick, converging tubes discernible on bottom (Fig. 406). Brown, sclerotized plate, not depicted, extends inwards along epigastric furrow, covering internal organs from inside. Yellow-brown, relatively very large spermathecae, with readily apparent thick, looping black ducts visible on dorsal view (Fig. 407).

Distribution: Mediterranean countries.

Israel: Along the Coastal Plain (8, 9), in mountainous parts (2, 11), near the Dead Sea (13) and in the sand dunes south to Be'er Sheva' (15).

Adults are found from March to June; some were observed near ants.

406

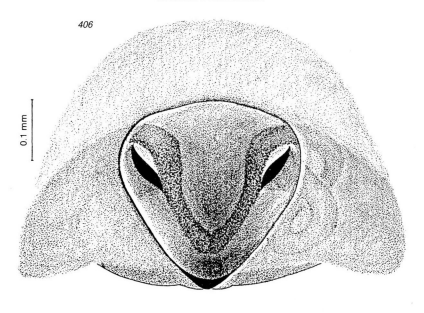

407

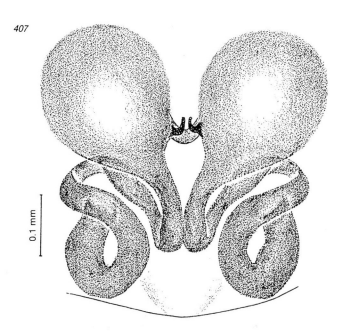

Figs 406–407: *Theridion uncinatum* Lucas, 1846; female
406. epigynum; 407. inner spermathecae, dorsal view

Theridion pustuliferum Levy & Amitai, 1982
Figs 408–410

Theridion pustuliferus Levy & Amitai, 1982a, *J. Zool. Lond.* 196:119.

Male unknown. Length of female 2.0–2.9 mm. Colour of carapace deep red-brown with black margins. Sternum reddish-black. Legs light, creamy, with black annulations and brown marks on articulations. Spherical opisthosoma with black, low posterior eminence, sometimes visible only in profile (Fig. 408). Dorsum intense white with four distinct black spots, arranged, two in a row, on each side; sometimes with additional median spots; hinder dorsal part of opisthosoma with a conspicuous black spot on low eminence and series of partly continuous black spots extending in a line to spinnerets; sides black mottled with white; venter dark with large, white patch below epigastric furrow and two small white spots obliquely placed, on sides of spinnerets.
Female Epigynum: Central depression on epigynal plate partly surrounded by thick, dark sclerotized, raised walls (Fig. 409); side loops of internal organs may sometimes be visible through integument (Fig. 409). Compact spermathecae yellow-brown with roughened surface (Fig. 410); ducts form thick, dark tubes coiling only once around base of spermathecae (Fig. 410).
Distribution: Israel.
Israel: Northern Galilee (1) and Judean Hills (11).
Adult females are found on plants in April. Spherical egg sac with a diameter of 2 mm contains about 30 eggs.

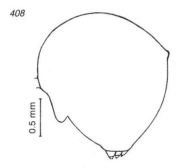

Fig. 408: *Theridion pustuliferum* Levy & Amitai, 1982; female,
opisthosoma, lateral view

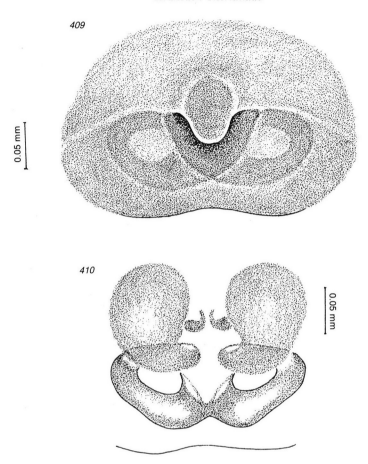

Figs 409–410: *Theridion pustuliferum* Levy & Amitai, 1982; female
409. epigynum; 410. inner spermathecae, dorsal view

APPENDIX

List of Emendations of Theridiidae in Israel or
Formerly Reported Mainly from Israel

Former name	Emendation
acuminatum (*Theridion*) Lucas, 1846. O.P.-Cambridge, 1872:281	*acuminata* (*Euryopis*) (Lucas, 1846)
albocinctum (*Theridion*) Lucas, 1846. O.P.-Cambridge, 1872:281	*albomaculata* (*Steatoda*) (DeGeer, 1778). Levy & Amitai, 1982b:15
albomaculata (*Aranea*) DeGeer, 1778	*albomaculata* (*Steatoda*) (DeGeer, 1778)
albomaculatus (*Lithyphantes*) (DeGeer, 1778). Bodenheimer, 1937:238	*albomaculata* (*Steatoda*) (DeGeer, 1778) Levy & Amitai, 1982b:15
ammonia (*Argyrodes*) Denis, 1947	*argyrodes* (*Argyrodes*) Walckenaer, 1841). Levy, 1985a:99
apicatum (*Theridion*) O.P.-Cambridge, 1872:281 Bodenheimer, 1937:237	*uncinatum* (*Theridion*) Lucas, 1846. Levy & Amitai, 1982a:116
argentea (*Teutana*) Caporiacco, 1933	*ephippiata* (*Steatoda*) (Thorell, 1875). Levy & Amitai, 1982b:22
argyrodes (*Linyphia*) Walckenaer, 1841	*argyrodes* (*Argyrodes*) (Walckenaer, 1841)
aulicum (*Theridium*) C.L. Koch, 1838. Strand, 1914:185; Bodenheimer, 1937:237	*aulicus* (*Anelosimus*) (C.L. Koch, 1838)
bifoveolatum (Theridion) Denis, 1944	*spinitarse* (*Theridion*) O.P.-Cambridge, 1876. Levy & Amitai, 1982a:84
braccatum (*Theridium*) C.L. Koch, 1841	*braccata* (*Dipoena*) (C.L. Koch, 1841)
conspicuum (*Theridion*) O.P.-Cambridge, 1872:285	*conspicua* (*Crustulina*) (O.P.-Cambridge, 1872)
dahli (*Lithyphantes*) Nosek, 1905	*dahli* (*Steatoda*) (Nosek, 1905)
denticulatum (*Theridion*) (Walckenaer, 1802). O.P.-Cambridge, 1872:280; Bodenheimer, 1937:237	*melanurum* (*Theridion*) Hahn, 1831
dialeucon (*Theridion*) Simon, 1890	*dialeucon* (*Anelosimus*) (Simon, 1890). Levy & Amitai, 1982a:124
epeirae (*Argyrodes*) Simon, 1866. O.P.-Cambridge, 1872:279	*argyrodes* (*Argyrodes*) (Walckenaer, 1841)
ephippiatus (*Lithyphantes*) Thorell, 1875	*ephippiata* (*Steatoda*) (Thorell, 1875)
episinoides (*Euryopis*) (Walckenaer, 1847). Roewer, 1942:450	*acuminata* (*Euryopis*) (Lucas, 1846). Levy & Amitai, 1981a:178
erigonforme (*Theridion*) O.P.-Cambridge, 1872:284	*erigoniformis* (*Steatoda*) (O.P.-Cambridge, 1872)
erigoniformis (*Asagenella*) (O.P.-Cambridge, 1872). Schenkel, 1937:238	*erigoniformis* (*Steatoda*) (O.P.-Cambridge, 1872)

Former name

erigoniformis (*Enoplognatha*) (O.P.-Cambridge 1872). Bodenheimer, 1937:238

femoralis (*Robertus*) (O.P.-Cambridge, 1872:291. Bodenheimer, 1937:238

flavomaculatum (*Theridion*) Lucas, 1846

flavomaculatum (*Theridion*) Lucas, 1846. O.P.-Cambridge, 1872:281; Bodenheimer, 1937:237

gerhardti (*Lithyphantes*) Wiehle, 1934. Bodenheimer, 1937:238

gibbosus (*Argyrodes*) (Lucas, 1846). Bodenheimer, 1937:237

grossa (*Teutana*) (C.L. Koch, 1838)

grossum (*Theridium*) C.L. Koch, 1838

hamatus (*Latrodectus*) (C.L. Koch, 1839). O.P.-Cambridge, 1870:819; 1872:288

hamatus (*Lithyphantes*) (C.L. Koch, 1839). O.P.-Cambridge, 1876:568

inscripta (*Euryopis*) (O.P.-Cambridge, 1872). Roewer, 1942:451; Bonnet, 1956:1824

inscriptum (*Theridion*) O.P.-Cambridge, 1872:284

latifasciatus (*Lithyphantes*) Simon, 1873

longicaudata (*Ariamne*) O.P.-Cambridge, 1872:277

longicaudata (*Rhomphaea*) (O.P.-Cambridge, 1872). Reimoser, 1919:28

lunatum (*Theridion*) (Clerck, 1757)

lunatus (*Araneus*) Clerck, 1757

lutipes (*luteipes*, laps.; *Theridion*) O.P.-Cambridge, 1869. O.P.-Cambridge, 1872:280

mactans hesperus (*Latrodectus*) Chamberlin & Ivie, 1935

mactans tredecimguttatus (*Latrodectus*) (Rossi, 1790). Levi, 1966:427

mandibulare (*Pachygnatha*) (Lucas, 1846). O.P.-Cambridge, 1872:294

mandibulare (*Theridion*) Lucas, 1846

Emendation

erigoniformis (*Steatoda*) (O.P.-Cambrige, 1872)

femoralis (*Brachycerasphora*) (O.P.-Cambridge, 1872) (a linyphiid, not a theridiid; Levy & Amitai, 1981b:45); Bosmans, 1994:233.

triangulosa (*Steatoda*) (Walckenaer, 1802) Levy & Amitai, 1982b:17

parathoracica (*Enoplognatha*) Levy & Amitai, 1981b:62 (along with an unidentified linyphiid)

maura (*Steatoda*) (Simon, 1909). Levy & Amitai, 1982b:21

argyrodes (*Argyrodes*) (Walckenaer, 1841)

grossa (*Steatoda*) (C.L. Koch, 1838)

grossa (*Steatoda*) (C.L. Koch, 1838)

paykulliana (*Steatoda*) (Walckenaer, 1806). Levy & Amitai, 1982b:19

paykulliana (*Steatoda*) (Walckenaer, 1806). Levy & Amitai, 1982b:20

nomen dubium (immature female *Dipoena* unidentifiable; Levy & Amitai, 1981a:178)

nomen dubium

latifasciata (*Steatoda*) (Simon, 1873)

longicaudata (*Argyrodes*) (O.P.-Cambrigde, 1872). Levy, 1985a:109

longicaudata (*Argyrodes*) (O.P.-Cambridge, 1872). Levy, 1985a:109

lunata (*Achaearanea*) (Clerck, 1757)

lunata (*Achaearanea*) (Clerck, 1757)

rufipes (*Theridion*) Lucas, 1846

hesperus (*Latrodectus*) Chamberlin & Ivie, 1935

tredecimguttatus (*Latrodectus*) (Rossi, 1790). Levy & Amitai, 1983:46

mandibularis (*Enoplognatha*) (Lucas, 1846) Levy & Amitai, 1981b:48

mandibularis (*Enoplognatha*) (Lucas, 1846)

Former name	Emendation
mandibularis (*Steatoda*) (Lucas, 1846). O.P.-Cambridge, 1876:568	*mandibularis* (*Enoplognatha*) (Lucas, 1846). Levy & Amitai, 1981b:48
maurus (*Lithyphantes*) Simon, 1909	*maura* (*Steatoda*) (Simon, 1909)
miami (*Theridion*) Levi, 1980	*melanostictum* (*Theridion*) O.P.-Cambridge, 1876. Levy, 1985a:114
ochraceus (*Lithyphantes*) Simon, 1908	*ephippiata* (*Steatoda*) (Simon, 1908). Levy & Amitai, 1982b:22
particeps (*Euryopis*) (O.P.-Cambridge, 1872). Bodenheimer, 1937:237	*sexalbomaculata* (*Euryopis*) (Lucas, 1846). Levy & Amitai, 1981a:180
particeps (*Theridion*) O.P.-Cambridge, 1872:282	*sexalbomaculata* (*Euryopis*) (Lucas, 1846). Levy & Amitai, 1981a:180
paykullianum (*Theridion*) Walckenaer, 1806	*paykulliana* (*Steatoda*) (Walckenaer, 1806)
paykullianus (*Lithyphantes*) (Walckenaer, 1806) Simon, 1892:82; Pavesi, 1895:7; Strand, 1914: 184; Bodenheimer, 1937:238	*paykulliana* (*Steatoda*) (Walckenaer, 1806). Levy & Amitai, 1982b:18
pictum (*Theridion*) (Walckenaer, 1802). Levy & Amitai, 1982:96.	*hemerobious* (*Theridion*) (Simon, 1914)
quadrimaculata (*Euryopis*) O.P.-Cambridge, 1876:569	*acuminata* (*Euryopis*) (Lucas, 1846). Levy & Amitai, 1981a:178
rufolineatum (*Theridion*) Lucas, 1846. O.P.-Cambridge, 1876:569	*aulicus* (*Anelosimus*) (C.L. Koch, 1838)
scripta (*Euryopis*) (O.P.-Cambridge, 1872). O.P.-Cambridge, 1876:569; Bodenheimer, 1937:237	*acuminata* (*Euryopis*) (Lucas, 1846). Levy & Amitai, 1981a:178
scriptum (*Theridion*) O.P.-Cambridge, 1872: 283	*acuminata* (*Euryopis*) (Lucas, 1846). Levy & Amitai, 1981a:178
septemmaculata (*Steatoda*) (Keyserling, 1884)	*erigoniformis* (*Steatoda*) (O.P.-Cambridge, 1872)
sexalbomaculatum (*Theridion*) Lucas, 1846	*sexalbomaculata* (*Euryopis*) (Lucas, 1846)
signata (*Asagena*) (O.P.-Cambridge, 1876). Strand, 1917:163	*erigoniformis* (*Steatoda*) (O.P.-Cambridge, 1872). Levy & Amitai, 1982b:26
signata (*Crustulina*) (O.P.-Cambridge, 1876). Simon, 1881:161	*erigoniformis* (*Steatoda*) (O.P.-Cambridge, 1872). Levy & Amitai, 1982b:26
signata (*Steatoda*) O.P.-Cambridge, 1876:568	*erigoniformis* (*Steatoda*) (O.P.-Cambridge, 1872). Levy & Amitai, 1982b:26
signatus (*Lithyphantes*) (O.P.-Cambridge, 1876). Simon, 1884:331	*erigoniformis* (*Steatoda*) (O.P.-Cambirdge, 1872). Levy & Amitai, 1982b:26
spirifer (*Theridion*) O.P.-Cambridge, 1863. O.P.-Cambridge, 1872:280	*aulicus* (*Anelosimus*) (C.L. Koch, 1838)
stictum (*Theridion*) O.P.-Cambridge, 1861	*sticta* (*Crustulina*) (O.P.-Cambridge, 1861)
tarsalis (*Euryopis*) Pavesi, 1875	*acuminata* (*Euryopis*) (Lucas, 1846). Levy & Amitai, 1981a:178

219

Former name	*Emendation*
trapezoidalis (Dipoena) Levy & Amitai, 1981a:184	*convexa (Dipoena)* (Blackwall, 1870)
tredecimguttata (13-*guttata*; XIII-*guttata*; *Aranea*) Rossi, 1790	*tredecimguttatus* (*Latrodectus*) (Rossi, 1790)
triangulosa (*Aranea*) Walckenaer, 1802	*triangulosa* (*Steatoda*) (Walckenaer, 1802)
triangulosa (*Teutana*) (Walckenaer, 1802). Simon, 1892:82; Strand, 1913:149, 1914:184; Bodenheimer, 1937:238	*triangulosa* (*Steatoda*) (Walckenaer, 1802). Levy & Amitai, 1982b:17
varians (*Theridion*) Hahn, 1831. O.P.-Cambridge, 1872:262; Bodenheimer, 1937:237	*hierichonticum* (*Theridion*) Levy & Amitai, 1982a:102
venustum (*Theridion*) Walckenaer, 1841. O.P.-Cambridge, 1872:281	*triangulosa* (*Steatoda*) (Walckenaer, 1802). (not *Enoplognatha ovata* (Clerck, 1757). Levy & Amitai, 1981b:48

Note

Theridion is neuter in Latin, hence original spellings were corrected by the Editor P. Merrett of N.I. Platnick's *Advances in Spider Taxonomy* 1981–1987, a supplement to Brignoli's *A Catalogue of the Araneae described between 1940 and 1981*. Manchester University Press, Manchester & New York, in association with the British Arachnological Society, 1989 (673 pp.).

Theridion accoensis Levy, 1985:117	= *T. accoense* Levy, 1985 (corrected in Platnick, 1989:203).
Theridion agaricographus Levy & Amitai, 1982:92	= *T. agaricographum* Levy & Amitai, 1982 (corrected in Platnick, 1989:203).
Theridion dafnensis Levy & Amitai, 1982:113	= *T. dafnense* Levy & Amitai, 1982 (corrected in Platnick, 1989:203).
Theridion gekkonicus Levy & Amitai, 1982:109	= *T. gekkonicum* Levy & Amitai, 1982 (corrected in Platnick, 1989:204).
Theridion hierichonticus Levy & Amitai, 1982:102	= *T. hierichonticum* Levy & Amitai, 1982 (corrected in Platnick, 1989:204).
Theridion jordanensis Levy & Amitai, 1982:102	= *T. jordanense* Levy & Amitai, 1982 (corrected in Platnick, 1989:204).
Theridion negebensis Levy & Amitai, 1982:106	= *T. negebense* Levy & Amitai, 1982 (corrected in Platnick, 1989:204).
Theridion ochreolus Levy & Amitai, 1982:91	= *T. ochreolum* Levy & Amitai, 1982 (corrected in Platnick, 1989:205).
Theridion vespertinus Levy, 1985:119	= *T. vespertinum* Levy, 1985 (corrected in Platnick, 1989:206).

REFERENCES*

Bodenheimer, F.S. (1937) 'Prodromus Faunae Palestinae', *Mémoires Inst. Égypte*, 33:1–286.

Bonnet, P. (1945–1961) *Bibliographia Araneorum*, Toulouse, Impre. Douladoure, 7 vols.

Bosmans, R. (1994) 'On some species described by O.P.-Cambridge in the genera *Erigone* and *Linyphia* from Egypt, Palestine and Syria (Araneae: Linyphiidae)', *Bull. Br. arachnal. soc.* 9(7): 233–235.

Bosmans, R., Vanuytven, H. & Van Keer, J. (1994) 'On two poorly known *Theridion* species collected in Belgium for the first time (Araneae: Theridiidae)', *Bull. Br. arachnol. Soc.* 9(7): 236–240.

Cambridge, O.P.- (1870) 'Notes on a collection of Arachnida made by J.K. Lord in the Peninsula of Sinai and on the African borders of the Red Sea', *Proc. zool. Soc. Lond.*, 1870: 818–823.

—— (1872) 'General list of the spiders of Palestine and Syria, with descriptions of numerous new species and characters of two new genera', *Proc. zool. Soc. Lond.*, 1872: 212–354.

—— (1876) 'Catalogue of a collection of spiders made in Egypt, with descriptions of new species and characters of a new genus', *Proc. zool. Soc. Lond.*, 1876: 541–630.

Levi, H.W.(1966) 'The three species of *Latrodectus* (Araneae) found in Israel', *J. Zool. Lond.*, 150:427–432.

Levy, G. & Amitai, P.(1979) 'The spider genus *Crustulina* (Araneae: Theridiidae) in Israel', *Israel J. Zool.*, 28:114–130.

—— (1981a) 'Spiders of the genera *Euryopis* and *Dipoena* (Araneae:Theridiidae) from Israel', *Bull. Br. arachnol. Soc.* 5(4):177–188.

—— (1981b) 'The spider genus *Enoplognatha* (Araneae: Theridiidae) in Israel', *Zool. J. Linn. Soc.*, 72: 43–67.

—— (1982a) 'The comb-footed spider genera *Theridion, Achaearanea* and *Anelosimus* of Israel (Araneae: Theridiidae)', *J. Zool. Lond.*, 196:81–131.

—— (1982b) 'The cobweb spider genus *Steatoda* (Araneae, Theridiidae) of Israel and Sinai', *Zool. Scr.*, 11:13–30.

—— (1983) 'Revision of the widow-spider genus *Latrodectus* (Araneae:Theridiidae) in Israel', *Zool. J. Linn. Soc.*, 77:39–63.

Levy, G.(1985a) 'Spiders of the genera *Episinus, Argyrodes* and *Coscinida* from Israel, with additional notes on *Theridion* (Araneae: Theridiidae)', *J. Zool. Lond.*, 207:87–123.

—— (1985b) *Fauna Palaestina, Arachnida II, Araneae: Thomisidae*. Israel Acad. Sci. Human., Jerusalem, pp. 1–116.

—— (1991) 'On some new and uncommon spiders from Israel (Araneae)', *Bull. Br. arachnol. Soc.* 8(7): 227–232.

---

* Listed are only selected references, notably on the theridiids of Israel and some referred to in the Appendix; complete lists of references are given in the publications by Levy and by Levy & Amitai.

Pavesi, P.(1895) 'Viaggio del Dott. E. Festa in Palestina, nel Libano e regioni vicine. XIV. Aracnidi', *Boll. Musei Zool. Anat. Comp. R. Univ. Torino*, 10(216):1–11.

Reimoser, E.(1919) 'Katalog der echten Spinnen (Araneae) des Paläarktischen Gebietes', *Abh. zool.-bot. Ges. Wien*, 10:1–280.

Roewer, C.F.(1942) *Katalog der Araneae*, Bremen, 1:1–1040.

Schenkel, E. (1937) 'Beschreibungen einiger afrikanischer Spinnen und Fundortsangaben', *Festschr. Strand*, 3:373–398.

Simon, E.(1881) *Les Arachnides de France*, Paris, 5(1):1–179.

——(1884) 'Etudes arachnologiques, 16ᵉ Mémoire. XXIII. Matériaux pour servir à la faune des Arachnides de la Grèce", *Annls Soc. ent. Fr.*, (6)4:305–306.

——(1892) 'Liste des arachnides recueillis en Syrie par M. le Dr. Théod. Barrois', *Rev. Biol. Nord Fr.*, Lille, 5:80–84.

Strand, E. (1913) 'Erste Mitteilung über Spinnen aus Palästina, gesammelt von Herrn Dr. J. Aharoni', *Arch. Naturgesch. Berlin*, 79(A,10):147–162.

——(1914) 'Zweite Mitteilung über Spinnen aus Palästina, gesammelt von Herrn Dr. J. Aharoni', *Arch. Naturgesch. Berlin*, 80(A,3):173–186.

—— (1917) 'Arachnologica varia. XIX–XX", *Arch. Naturgesch. Berlin*, 82(A,2):158–167.

Vanuytven, H., Van Keer, J. & Poot, P. (1994) 'Kogelspinnen verzameld in Zuid-Europa door P. Poot (Araneae, Theridiidae)', *Nwsbr. Belg. Arachnol. Ver. 9*(1): 1–19.

INDEX

Synonyms in italics. The principle reference to each valid name in bold type.

223

Geographical Areas in Israel and Sinai

KEY

1. Upper Galilee
2. Lower Galilee
3. Carmel Ridge
4. Northern Coastal Plain
5. Valley of Yizre'el
6. Samaria
7. Jordan Valley and Southern Golan
8. Central Coastal Plain
9. Southern Coastal Plain
10. Foothills of Judea
11. Judean Hills
12. Judean Desert
13. Dead Sea Area
14. 'Arava Valley
15. Northern Negev
16. Southern Negev
17. Central Negev
18. Golan Heights
19. Mount Hermon
20. Northern Sinai
21. Central Sinai Foothills
22. Sinai Mountains
23. Southwestern Sinai

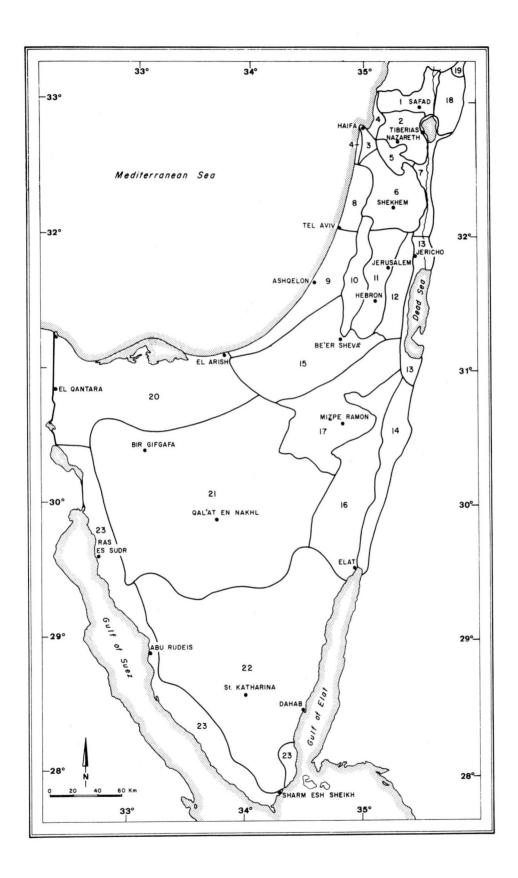